The Little Book
of Science

John Gribbin

The Little Book of Science

BARNES
&NOBLE
BOOKS
NEW YORK

This edition published by Barnes & Noble, Inc.,
by arrangement with Penguin Books Ltd.

1999 Barnes & Noble Books

ISBN 0-7607-1687-0

Printed and bound in the United States of America

99 00 01 02 03 MP 9 8 7 6 5 4 3 2 1

OPM

Contents

atom

If, in some cataclysm, all of scientific knowledge were to be destroyed, and only one sentence passed on to the next generations of creatures, what statement would contain the most information in the fewest words? I believe it is the *atomic hypothesis* (or the atomic *fact*, or whatever you wish to call it) that *all things are made of atoms – little particles that move around in perpetual motion, attracting each other when they are a little distance apart, but repelling upon being squeezed into one another*. In that one sentence, you will see, there is an *enormous* amount of information about the world, if just a little imagination and thinking are applied.

Richard Feynman,
Six Easy Pieces, 1998

atom of heredity

Genes are so important in evolution that they have sometimes been called the 'atoms of heredity' – they are the basic units involved in passing on biological information from one generation to the next. Genes specify particular things about the kind of body (phenotype) they produce – for example, there is a gene for eye colour. In species that reproduce sexually, like people, each individual inherits one copy of every gene from each parent. The whole set of genes is called the genotype; each cell of the body contains a complete set of genes – a complete genotype. The different versions of a particular gene are called alleles, and the genotype of a body might include one allele for blue eyes and one for brown eyes. In that case, the body will have brown eyes, because that allele is dominant; a person only has blue eyes if both copies of the eye-colour gene they inherit specify blue eyes. The blue-eye allele is said to be recessive.

In most cases there are far more than just two alleles to choose from, providing an enormous variety of genotypes in the population at large. Any allele carried by any woman might, in principle, be paired in the next generation with any allele for the same gene carried by any man. Each individual has only one pair of alleles, but there are many more in other human bodies – the gene pool. This is the raw material for evolution.

Imagine a change in the Sun's light, in the pre-civilization era, which made it so difficult for people with brown eyes to see that they could not find food. In a very few generations, the allele for brown eyes would disappear from the gene pool, because people with brown eyes would not leave descendants. Selection operates on individuals, but the effects are seen in the population at large.

The mathematics describing how small variations cause big changes was developed in the 1920s. R. A. Fisher showed that if a new allele, produced by mutation from an old one, gives the animal or plant that possesses it just a 1 per cent advantage, in terms of Darwinian fitness, over those that do not, the allele will spread through the entire population within a hundred generations. An advantage too slight even to be noticed by human observers monitoring a wild population can ensure the success of a mutated gene.

biological cell

The cell is the basic unit of life. A complex organism like a human being is made up of a huge number of cells – about a hundred thousand billion cells in your body, which is about a thousand cells for every bright star in our Milky Way Galaxy. But each individual cell has all the attributes of life, including the ability to reproduce itself. Most important of all, the fertilized egg of an animal, or the seed of a plant, is itself a single cell which develops into the adult form by a repeated process of cell division.

There are about 256 different kinds of cell in your body, all made from a single fertilized egg cell by repeated division. But because of the power of repeated doubling, it only takes about fifty cell divisions to make all the cells in your body – 2^{50} is about 10^{15} (one million billion) individual cells.

The kind of cell which makes up a human body has a well-developed central region, called the nucleus (when Ernest Rutherford chose the name 'nucleus' for the central part of an atom, he did so in a conscious imitation of the nomenclature used in cell biology). Such cells are said to be eukaryotic. Bacteria have a less well-developed cell structure with no nucleus; such cells are called prokaryotic, and represent an earlier stage in evolution.

The nucleus, contained within a membrane within the cell itself, is the region where DNA is stored in a tightly coiled up form. The rest of the cell, outside the nucleus, is called the cytoplasm, and is the factory of the cell, the region where the biological machinery of the cell puts together various molecules essential to the well-being of the organism, using instructions coded in the DNA to carry out chemical reactions in the right sequence.

This chemical activity takes place in a jelly-like material called cytosol. The entire cell is enclosed within an outer membrane, which restricts the flow of chemicals into and out of the cell, but is not a completely impervious barrier. The cells in a human body are each about 20 micrometres (20 millionths of a metre) across, so it would take ten thousand of them to cover the head of a pin. But there are three times as many cells in your cerebral cortex alone (Hercule Poirot's 'little grey cells') as there are people alive on Earth at the end of the twentieth century.

Bishop Berkeley's bucket

If you hang a bucket of water from a long twisted rope, and let it start to spin as the rope untwists, at first the bucket rotates but the water stays still. Gradually, friction makes the water rotate, and as it does so it forms a concave surface, produced by centrifugal force. Now grab the bucket and stop it moving. At first, the water keeps moving, and stays concave. Then, it slows down and stops, making a flat surface. How does the water 'know' when it is rotating, and ought to have a concave surface? It certainly doesn't measure its rotation relative to the bucket, because it can be concave when both bucket and water are moving, or when the water is moving and the bucket is not.

This highlights the puzzle of inertia, which the Irish philosopher, mathematician and bishop, George Berkeley, explained in the early eighteenth century by saying that objects measure their motion (somehow) relative to the most distant objects in the Universe. He referred to the distant stars, but today we would think in terms of distant galaxies.

The idea was developed further by Ernst Mach in the 1860s, and is often known as Mach's principle. It is still a deep mystery how it works. Einstein tried to construct his general theory of relativity to incorporate Mach's principle, but didn't quite succeed. Even so, there seems little doubt that when I try to push a child on a swing, and have to use energy to make it start to move, somehow the influence of all the matter in the Universe, in all the stars and galaxies, resists my efforts and provides the inertia that I have to do work against. The coffee in your cup rises up the side of the cup when you stir it because it knows that the distant galaxies are there.

chaos

In everyday life, chaos is a state of complete disorder in which the normal laws of cause and effect seem to be abandoned. The intriguing thing about the phenomenon scientists call chaos is that it involves completely unpredictable behaviour which follows from absolutely predictable (deterministic) laws. If you knew the position of every particle in the Universe, and how they were moving within it, at any instant of time, then according to Newton's laws you would be able to predict the future and reconstruct the past of the entire Universe. But this is easier said than done. Specifying the position of even a single particle precisely, may involve numbers – like the pi (π) which is associated with circles – which have an infinite number of non-repeating digits after the decimal point (irrational numbers). So to specify the state of just one particle *precisely* may involve an infinitely big computer.

This wouldn't matter if small errors in the calculations stayed small. But chaos results when a tiny difference in starting conditions grows enormously as you calculate the future. A simple example is a raindrop falling on a tall, sharply edged mountain ridge. Somewhere in the Rockies, if the drop falls on one side of the ridge it will eventually join a river flowing into the Pacific Ocean. If it falls a tiny bit to the east, it will join a river flowing into the Atlantic Ocean.

The classic example of chaos at work is in the weather. If you could measure the positions and motions of all the atoms in the air at once, you could predict the weather perfectly. But computer simulations show that tiny differences in starting conditions build up over about a week to give wildly different forecasts. So weather predicting will never be any good for forecasts more than a few days ahead, no matter how big (in terms of memory) and fast computers get to be in the future. The only computer that can simulate the weather is the weather; and the only computer that can simulate the Universe is the Universe.

cosmic string

Some models of the way the Universe emerged from the Big Bang imply that material left over from the Big Bang itself might have got frozen into long, thin tubes and trapped. This material is called cosmic string, and although nobody has yet found proof that it exists, physicists have worked out what kind of properties it ought to have, if it is ever found.

The tube containing stuff from the Big Bang would be far finer than the finest hair – it would take a hundred thousand billion of these tubes laid side by side to stretch across the diameter of a single atomic nucleus. But even though it is so narrow, each centimetre of the string would contain the equivalent of ten thousand billion tonnes of ordinary matter. So a loop of cosmic string with a circumference of just 1 metre would weigh as much as the whole Earth.

If cosmic string does exist, it must be in the form of loops, because if it had ends the frozen Big Bang stuff inside would leak out. The extreme energy of the stuff inside the strings would make them taut, trying to expand, making tension like a stretched elastic band. They would ripple and vibrate, twanging like plucked guitar strings, as they made their way across the Universe. This twanging would drain energy out of the loops, turning their mass into pulses of energy (gravitational radiation) and draining them away, eventually, to nothing.

In one speculative, but intriguing, scenario, such supermassive loops of string, light years across, could have been left over from the Big Bang. The intense gravitational pull of the loops would have attracted huge clouds of gas, which formed the galaxies that we see today. But, by the time the galaxies formed, the twanging of the loops would have shrivelled them up, leaving nothing behind. The galaxies we see today could be like the grin on the face of the Cheshire cat, while its body (the cosmic string) has faded away.

DNA

The molecule of life which carries all the information needed for the care and maintenance of an organism (anything from a bacterium to a human being) is called DNA, short for deoxyribose nucleic acid. Molecules of DNA are generally very long chains, made up of many components linked together. The backbone of each chain is a repeating sequence of alternating phosphate groups and sugars. But what matters for life is that each of the sugar units in the chain has one of four possible different chemical units sticking out sideways from the chain. These molecules (called bases) are adenine (A), cytosine (C), guanine (G) and thymine (T). Two such chains 'hold hands', with A always linking up with T, and C always linking up with G, to make the famous double helix. But what matters in terms of information storage is the sequence of 'letters' along each chain – AGCCATGTCATT… or whatever it might be. This works like a message in a four letter alphabet – a message 2500 million letters long in the case of human beings, equivalent to an ordinary book a million pages long. This is ample to describe completely how to build and run a human body.

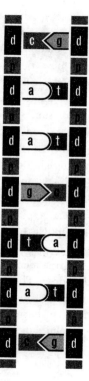

key

t	- thymine
a	- adenine
g	- guanine
c	- cytosine
d	- deoxyribose
p	- phosphate

13

drug-resistant bacteria

Even among scientists, people who have not specialized in biology sometimes fail to understand the way evolution works. Chemists who invented new pesticides seem to have been astonished when insects developed resistance to them, and recently there have been many scare stories about 'superbugs', bacteria which are resistant to penicillin and other drugs which used to be highly effective against the same kind of bacteria. To a biologist, though, this is simply an example of evolution at work.

When drugs are used to treat a particular disease, they kill off all the susceptible bacteria, and the patient recovers. But, by definition, if there are any bacteria that survive the treatment, they are the ones most resistant to that drug. The more you kill the bacteria that are susceptible, the more opportunity you give to the resistant bacteria to breed, and to pass on their genetic qualities that make them resistant to the drug. Bacteria breed far more quickly than human beings, so it is no surprise to an evolutionary biologist that in a matter of years or decades all of the bacteria that cause that particular disease have evolved resistance to that particular drug.

This is why doctors are now advised not to prescribe antibiotics unless really necessary, to delay as long as possible the inevitable evolution of resistant bugs. And it is also why you should finish a course of treatment even if you start to feel better, to finish off any remaining, partly resistant bacteria in your system.

There is, though, a delicious irony in this tale. Cotton farmers in the southern states of the US, where resistance to Darwinian ideas is at its strongest, are struggling every year with the consequences of evolution at work in their own fields, where insect pests are increasingly unconcerned by traditional pesticides.

electromagnetic spectrum

Visible light forms only a small part of an electromagnetic spectrum that stretches from very long-wave radio waves to highly energetic gamma rays. The only difference across the spectrum is the length of the waves involved. Our eyes are only sensitive to the part of the spectrum where the Sun's radiation is strongest – which is no coincidence, because eyes have evolved to make use of the radiation that is available. As well as visible light, some radio waves also penetrate the atmosphere of the Earth; but most of the electromagnetic spectrum is blocked by the atmosphere, which is why many astronomical observations are carried out using satellites, above the atmosphere.

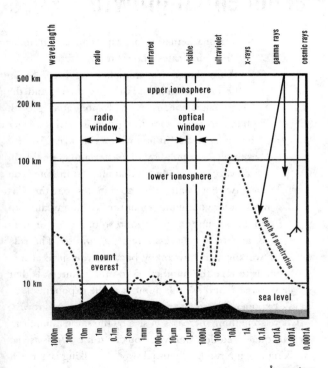

Å=angstrom
μm=micrometre

exponential growth

In mathematics, an exponent is a number that indicates how many times something should be multiplied by itself. So in the expression x^2, 2 is the exponent, and the expression means $x \times x$. Exponential growth happens when something that is getting bigger increases in size in line with a rule like this – for example, a population of bacteria may double in size every 10 minutes, so that after 10 minutes you have twice as many, after 20 minutes four times as many, after 30 minutes eight times as many, after 40 minutes 16 times as many, and so on. The numbers 4, 8, and 16 are successively 2^2, 2^3 and 2^4. The rule for working out how many bacteria you have at any time is to take the number of 10-minute intervals that have passed, and use this number as the exponent of 2.

Exponential growth contrasts with linear growth, where something increases in size by the same amount in each equal-sized time interval. You can think of this as like walking down a road. The usual way of walking is linear – each pace might be, say, 1 metre long. After 10 paces, you have covered a distance of 10 metres. But suppose you had super-seven-league boots, so that each pace you took was twice as long as the one before. The first pace is still 1 metre, the second 2 metres, the third 4 metres, and so on. Every subsequent step is bigger than all the previous steps put together, and the eleventh step alone will cover 2^{10} metres, which works out at just over a kilometre.

This kind of exponential growth is important in the theory of the Big Bang, where it describes the very rapid growth of the early Universe, called inflation, in which the size of the Universe doubled a hundred times in a split second, taking a region 10^{20} times smaller than a proton and inflating it into a sphere roughly the size of a grapefruit. Inflation then ended, and linear growth took over, leaving a hot expanding fireball that developed into the Universe we see today. Exponential growth is also important in chaos theory, and describes the way errors grow wildly out of control in non-linear systems, and make it impossible, for example, to predict the weather accurately more than a few days ahead.

Feynman diagram

In the world of particle physics, the most useful tool of the theorists is a kind of drawing known as a Feynman diagram, named after its inventor, Richard Feynman.

These diagrams are deceptively simple, and at one level they can be regarded as a kind of cartoon representation of how particles interact with one another. In the classic Feynman diagram shown here, an electron moving through space exchanges a photon (a particle of electromagnetic radiation) with another electron, and the two electrons recoil from one another as a result.

But in Feynman's own words, 'the diagrams were intended to represent physical processes *and the mathematical expressions used to describe them*' (my italics). In particular, each vertex on the diagram is a geometrical shorthand notation for a series of mathematical equations, which may be quite complex.

This is similar to the way in which a single smooth curve on a graph may represent a mathematical (algebraic)

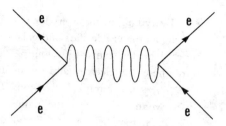

Feynman diagram

expression. A parabola, for example, can be represented by a line on a graph, or by the equation $y = x^2$. Just as an expert can 'read' a graph and tell you the relevant equation, so an expert in particle theory can 'read' a Feynman diagram and translate it into the appropriate mathematical expression. The advantage of using the diagram is that the 'output' from an interaction, or a series of interactions, can be read off in this way without actually carrying out all the steps in the mathematical calculation of how the particles interact.

fundamental particles

Everything in the everyday world (including all the visible stars and galaxies) can be explained in terms of just four particles. Two of these, the electron and the neutrino, are members of a family called leptons. The other two, whimsically given the names 'up' and 'down', are members of a family called quarks.

Two up quarks and one down quark together make up a proton; two down quarks and one up quark together make up a neutron. Protons and neutrons together make up the nuclei of atoms, with electrons forming a cloud around the nucleus. The neutrino is involved in the process of radioactive decay.

For reasons nobody understands, this basic plan is triplicated in nature. In high-energy particle experiments, it is possible to manufacture a second generation of all these particles, identical to the first generation except for their mass. These are the quarks known as 'charm' and 'strange', a heavy electron called the muon, and a neutrino associated with the muon. Finally, there is an even heavier set of particles – the 'top' and 'bottom' quarks, the tau particle, and its neutrino. But although these particles can be manufactured in particle accelerators, they are unstable and quickly decay into their stable, lightest counterparts.

The neat symmetry in this pattern, with three pairs of quarks and three pairs of leptons, hints at some deep truth about the physical world, but it is not yet clear what this is.

Leptons	Quarks
e^-	u
ν_e	d
μ^-	c
ν_μ	s
τ^-	t
ν_τ	b

Gaia

A scientific idea that has (perhaps appropriately) taken on a life of its own in recent decades is Jim Lovelock's concept of Gaia, which sees the Earth as a self-regulating system in which conditions suitable for life are maintained by feedback processes involving both living things and the non-living part of the planet. Alas for New Agers and mystics, though, this does not mean (and Lovelock never said it did) that the planet itself is 'alive'; the name Gaia, from the Greek Earth-goddess, is just a nifty name, not to be taken literally.

But the scientific power of Lovelock's reasoning is clear if you imagine a visitor from another star studying the inner planets of the Solar System through a telescope as they approach. Venus and Mars, the two nearest neighbours to the Earth, both have stable atmospheres in chemical equilibrium, rich in the unreactive gas carbon dioxide. But Earth is odd. It has a highly reactive atmosphere, far from chemical equilibrium, in which gases such as methane and oxygen mix together. They are so reactive that we actually use methane ('natural gas') in cooking stoves, burning it in oxygen from the air. The two gases coexist, at more or less the same composition for long periods of time, because they are being produced by biological processes as rapidly as they are being destroyed in chemical reactions.

Our hypothetical aliens would immediately head for Earth, knowing that there was no point in looking for life on Mars or Venus – a lesson which, unfortunately, NASA has yet to take heed of.

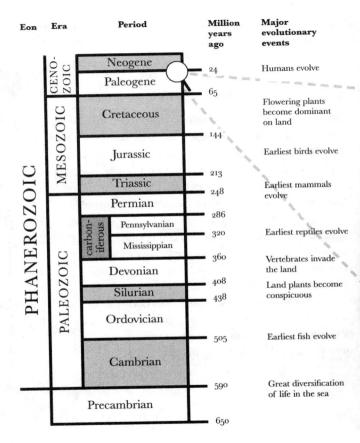

Eon	Era	Period		Million years ago	Major evolutionary events
PHANEROZOIC	CENO-ZOIC	Neogene		24	Humans evolve
		Paleogene		65	
	MESOZOIC	Cretaceous		144	Flowering plants become dominant on land
		Jurassic		213	Earliest birds evolve
		Triassic		248	Earliest mammals evolve
	PALEOZOIC	Permian		286	
		carbon-iferous	Pennsylvanian	320	Earliest reptiles evolve
			Mississippian	360	Vertebrates invade the land
		Devonian		408	Land plants become conspicuous
		Silurian		438	
		Ordovician		505	Earliest fish evolve
		Cambrian		590	Great diversification of life in the sea
		Precambrian		650	

geological timescale

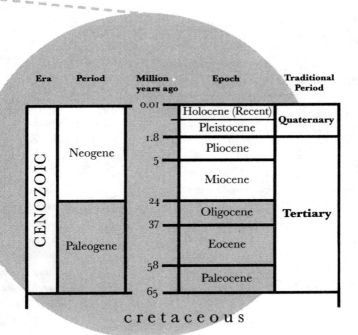

human genome project

Chemical techniques for analysing strands of DNA are now so sophisticated that a project to 'map' the entire human genome is well underway, and should be completed early in the twenty-first century. This is the first step to identifying and repairing faulty genes that cause hereditary diseases (such as cystic fibrosis) and then to the more science-fiction-like possibility of tailoring genes to produce individuals with characteristics that are regarded as desirable.

The entire genetic code of a human being is stored on strands of DNA (actually double strands coiled around each other in the famous double helix) that would stretch over a distance of about two metres, if they were pulled out tight and laid end to end like railway lines. The DNA sequence is made up of about three billion consecutive 'letters', each denoted by one of four chemical units known as bases, in order along the railway lines (each base is mirrored by a counterpart on the other track, so a letter in genetic code corresponds to a DNA base pair). The human genome project is the most ambitious piece of biological research ever undertaken, involving many research groups around the world and costing tens of billions of dollars; the total time required to measure the positions of the three billion base pairs in the sequence is estimated as 30,000 person years, since it takes one person one year to 'sequence' a piece of DNA 100,000 bases long.

The first human gene to be sequenced (in 1983, two years before the genome project got underway) was the one that causes Huntington's disease. In 1987, the gene that causes muscular dystrophy was mapped, and the gene that causes cystic fibrosis in 1989. In the 1990s, biologists were ticking off their list genes responsible for many mundane properties of the human body, although the project largely faded from the headlines as the work became routine. The ethical implications of this work are enormous, though, and this is likely to lead to heated debates when people realize how close the project is to completion.

human origins

People are apes. We share 99 per cent of our genetic material, DNA, with the African apes – the chimpanzee and the gorilla. Studies of the biological molecules, including DNA, in many species have enabled biologists to work out how long it has taken for this small difference between humans, chimps and gorillas to accumulate. The evidence shows that the three species shared a common ancestor as recently as 5 million years ago, and that after the gorilla line split off from the common stock, the chimps and ourselves shared a common ancestry for a little longer, perhaps until only 4 million years ago. You are *more* closely related to a chimp than the chimp is to a gorilla.

One of the most intriguing features of this discovery is that it means that the common ancestor of ourselves and the chimps (at least – possibly the gorillas too) had developed the ability to walk upright. This is clear from the evidence of fossils found in East Africa. The other apes seem to have largely abandoned this ability, and evolved to fit lifestyles suited for forest dwellers, while our ancestors retained and refined the ability to walk upright, and only at a later stage developed the intelligence which we regard as the supreme characteristic of human beings.

The fossil evidence shows that these key developments in human evolution did happen in East Africa, and it is widely accepted that they were triggered by climate changes. A series of Ice Ages swept over the Northern hemisphere in the past 5 million years, and the forests of East Africa were hit by drought every time the Earth cooled (because when the world is cool, less water evaporates from the oceans to make rain). These waves of drought, which made the forests shrink, probably forced our ancestors to adapt to a life on the plains, and put intelligence at a premium. The only other evolutionary alternative was to become increasingly skilful at life in the dwindling forests, like the chimp and gorilla. So we are descended from the unsuccessful apes, who got pushed out from the forests and had to find a new way to make a living.

inside the atom

Atoms are almost big enough to comprehend. Atoms are the smallest part of an element, the unit that takes part in chemical reactions, with atoms joining together to make molecules. Atoms are much the same size as one another, with the largest, caesium (Cs), about 0.0000005 mm (that is, 5×10^{-7} mm) across. So it would take 10 million caesium atoms side by side to stretch across the gap between two of the points on the serrated edge of a postage stamp.

But it is when you try to probe inside the atom that the scales involved become almost incomprehensible. An atom is made up of a tiny central nucleus, which carries positive electric charge, surrounded by a cloud of electrons, which carry negative electric charge (so overall the charge on an atom is zero). The electron cloud forms the visible face of the atom, which controls its chemical behaviour, in line with the rules of quantum physics. Atoms join up to make molecules, forming chemical bonds, when some electrons are shared between atoms, forming a kind of bridge between them; this bonding occurs only if the arrangement of electrons in the molecule is more stable (less energetic) than if the atoms remained unjoined.

But the electrons carry less than 0.025 per cent of the mass of the atom, even though they fill most of its volume and determine its chemical nature. The nucleus has almost all the mass, but occupies only a tiny volume, about 10^{-12} mm across. In round terms, the whole atom is one hundred thousand times bigger than the nucleus, and if the nucleus were actually a few millimetres across the whole atom would just fit into a football stadium.

This gives you a feel for just how empty 'solid' matter really is. The solid bits (nuclei) are like peas, each at the centre of a volume as big as a football stadium. Things like the keyboard on my computer, or the pages of the book you are reading, only seem solid because these solid nuggets are linked by a web of forces and quantum fields, which lock the nuclei in place. When you press your finger against a rock, the only actual contact is between clouds of electrons in the atoms in your fingertips, pushing against clouds of electrons in the atoms in the surface of the rock.

interactions

The way in which the fundamental particles affect each other is described by four interactions. These are sometimes called the forces of nature, but interactions is a better term because one, the weak interaction, doesn't behave like a force in the everyday sense of the word.

The strongest interaction is called, logically enough, the strong interaction. It is the force that holds atomic nuclei together, even though nuclei carry positive electric charge, which is trying to blow them apart. The strong interaction only has a short range – about the same as the size of an atomic nucleus, which is why nuclei are that size.

The electromagnetic interaction is the next strongest force, about 1/137 as strong as the strong force. So if you have 137 protons in one place, the electric repulsion will overcome the ability of the strong interaction to hold them together. In fact, because the protons are spread out over a small volume, not literally in the same place, the strong force cannot hold quite that many together, and the stable element whose atoms have most protons in their nuclei is uranium, with 92.

The next weakest of the four interactions is the weak interaction.

This doesn't act like a force, but its strength is regarded as being about one ten-thousandth of a billionth (10^{-13}) of the strength of the strong force. The weak force makes isolated neutrons turn into protons, each emitting an electron and a neutrino (strictly speaking, an antineutrino, the anti-matter counterpart to a neutrino) as it does so. It is also responsible for nuclear fission.

The weakest interaction of them all is gravity, which has a strength of only 10^{-38} compared with the strong force. Gravity only seems strong to us because its effects add up, and it has a long range. But it takes the gravity of every particle in planet Earth to hold you down. Electromagnetism also has a long range, but its effects don't add up because every negative charge is balanced by a positive charge, and every north pole is balanced by a south pole. Even so, you can see the relative strength of the two forces by noticing that it takes the gravity of the entire Earth to break the electromagnetic bonds between a few atoms in the stalk of an apple and pull the apple off a tree.

jumping genes

Wild maize (sweetcorn) doesn't have the even rows of yellow seeds of the kind you see in the supermarket. Instead, the seeds are several different colours, seemingly dotted around at random. In the 1940s, the American geneticist Barbara McClintock found that the genes which fix the colour of a particular seed in the cob could move about within the genome, jumping from one chromosome to another.

Similar jumping genes were later (in the 1970s) found to exist in bacteria, where a piece of genetic material that makes bacteria resistant to antibiotics could move from one part of the genome to another. It became clear that in this case the genes themselves do not really 'jump', but make copies of themselves which are then inserted into other chromosomes (or similar loops of DNA called plasmids); they are now known as transposons.

The ability of genes to move around within the genome makes species evolve more rapidly, which can, under some circumstances, confer an evolutionary advantage. And the spread of these jumping genes is encouraged in some bacteria because they undergo a kind of sexual reproduction, in which two bacteria link up and exchange genetic material before going on their separate ways.

It is even possible that bits of genetic material may get transferred from one species to another, perhaps by the action of viruses. This is a major reason why many scientists are concerned about the release of genetically engineered crops into the fields – if, for example, a food crop has been modified to make it resistant to weedkillers, the genes coding for that resistance may be able to 'jump' into weeds themselves, making them that much harder to control.

Jupiter effect

Don't believe everything you read in books. In the mid 1970s, two young astronomers came up with the idea that earthquakes might be triggered by changes in solar activity, which could in turn be triggered by the influence of planetary alignments on the Sun. They predicted that a major earthquake would hit in California as a result, in 1982 when the planets formed a tight group on one side of the Sun. This became known as the 'Jupiter effect', from the name of the largest planet in our Solar System.

But they were wrong. The Sun did not behave in an unusual way at the specified time, and the predicted earthquake never happened. The idea had been tested by observation and experiment and proved wrong, so it was abandoned by its founders. That should have been the last we heard of the Jupiter effect.

Unfortunately, by that time a garbled version of the idea had been picked up by astrologers and millennial doom-sayers, who noticed that another tight grouping of the planets would occur at the end of the twentieth century, ushering in the new millennium. Ignoring the fact that the 'effect' had been disproved, they predicted not just one earthquake but a wave of global catastrophes occurring as a result, alarming a lot of gullible people needlessly.

But don't worry. There is no Jupiter effect, and the world will not come to an end in 1999 or in the year 2000. How do I know? Because I was one of those young astronomers who made the original 1982 prediction. And as Erwin Schrödinger once said of the quantum theory he helped to found, I don't like it, and I'm sorry I ever had anything to do with it.

KT event

It has now been established beyond reasonable doubt that the 'Doomsday comet' that brought an end to the era of the dinosaurs struck the region that is now the Yucatan peninsula of Mexico, some 65 million years ago. The impact was caused by an object about 10 kilometres across, which produced a crater 180 kilometres wide. The energy released in the impact was simply the kinetic energy (energy of motion) of this lump of cosmic debris, being brought to a sudden halt from a speed of about 50 kilometres per second. The energy released in this way was equivalent to the explosion of a thousand million megatons of TNT. It is just the same process as the one that warms the brakes on your car when it comes to a halt – only more so.

While the crater itself seethed with molten rock, fragments from the impact would have been thrown out on ballistic trajectories, re-entering the Earth's atmosphere far away, like meteorites in their own right. The overall effect would have been to heat the entire surface of the Earth by 10 kilowatts per square metre for several hours after the impact. Animals out in the open at the time would have been roasted alive, while vegetable matter burned, creating a global soot layer that is a distinctive feature in the geological record at the boundary between the Cretaceous and Tertiary geological periods (this is known as the 'KT event', not the 'CT event', to avoid confusion with the Cambrian period).

After the heat came the cold. Smoke and debris in the atmosphere would have blotted out the Sun for months. And as the atmosphere cleared, acid rain and poisonous mutagenic material fell to Earth and entered the food chain. It is small wonder that 70 per cent of all species alive on Earth at the time were wiped out at the end of the Cretaceous. The good news is that by clearing rivals out of the way, this catastrophe opened the way for mammals to diversify and evolve, eventually producing ourselves. The bad news is that the 'terminal Cretaceous event' was not an isolated incident. Such impacts have happened before, and will happen again.

Kuiper belt

One of the most dramatic, but largely unsung, astronomical discoveries of the 1990s is a belt of cometary debris orbiting beyond the orbit of Neptune, the most distant large planet in our Solar System. This is known as the Kuiper Belt, after the astronomer Gerard Kuiper, who first predicted the existence of the belt on theoretical grounds back in 1951.

The discovery of comets in the Kuiper Belt had to await improvements in telescope technology, so the first such object, dubbed QB1, was only detected in 1992. It is about 200 km across (ordinary comets, like Halley's Comet or Hale–Bopp, are typically only 10 km across), and orbits 44 times further out from the Sun than we are. By 1995, twenty such objects had been identified, and they are so difficult to find even with modern equipment that this suggests that there are at least 35,000 icy objects, each at least 100 km across, orbiting the Solar System just beyond the orbit of Neptune. To put that number in perspective, there are only 200 objects that big in the much more famous asteroid belt that lies between the orbits of Mars and Jupiter. The total mass of these objects is about 3 per cent of the mass of the Earth, and they are probably accompanied by between one billion and ten billion objects the size of Halley's Comet.

Computer simulations show that from time to time one of these objects will be disturbed by the gravity of the outer planets into an orbit which takes it diving into the inner Solar System, close to the Earth. Such a supercomet will then break up in the inner Solar System, scattering dusty debris in a temporary ring around the Sun which might trigger an Ice Age on Earth, by blocking out some of the Sun's heat. On very rare occasions, bits of cometary debris may even strike the Earth – the disaster that brought an end to the era of the dinosaurs, some 65 million years ago, fits the profile of such an event.

left-handed Universe

Our Universe is left-handed. This shows up only at a very subtle level, in particle interactions that involve the weak interaction, the force of nature that is responsible for, among other things, radioactive decay. Physicists describe this asymmetry in the physical world in terms of a property they call parity, or P, which can be understood in terms of the way things would look if they were reflected in a mirror.

There is a particular kind of particle, called the muon, which decays, as a result of the weak interaction, and spits out electrons as it does so. Muons also possess the property called spin, and rotate as they move through space, producing a corkscrew-like motion. If a corkscrew is reflected in a mirror, a right-handed screw becomes a left-handed screw, and vice versa. If the muon were to radiate electrons in both directions when it decayed, forwards and backwards along its line of flight, the mirror image would look just the same, because although the handedness of the corkscrew would be reversed, there would still be electrons

44

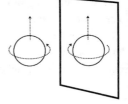

diagram 1. 'handedness' of a spinning particle moving at constant velocity reversed in a mirror

emerging at each end, some in the direction of its motion and some in the opposite direction. But in fact the advancing muon only ejects electrons in the direction of its motion. The mirror image of this would be a muon rotating in the opposite sense (with the opposite handedness) but still emitting electrons in the direction of its motion. No such particle decay occurs, so there is no true mirror image of the process of muon decay; the opposite version of muon decay that actually occurs in nature is like the version I have just described turned upside down, so that the electrons still emerge in the same direction relative to the spin of the muon. The weak force cannot be reflected in a mirror – or in technical terms, parity is violated.

45

The way in which muons decay provides an absolute standard of left and right, so we could explain what we mean by right-handed and left-handed in a radio message to aliens far away in space by describing the process of muon decay. In this sense, the Universe is left-handed – but nobody knows why.

electrons

mirror image is **not** seen

this kind of decay **is** seen

life

Life is a self-organizing complex phenomenon that occurs (sometimes) in places where energy is flowing from a hotter object to a cooler object. It can never occur without such an energy flow, which is an essential requirement for life (but not the only requirement) because the flow of energy is necessary in order to create a local region in which the second law of thermodynamics can temporarily be held at bay. This means that order can appear out of a disordered state in this bubble, even though in the Universe at large more than enough disorder is created to cancel out this growth of local order.

The place where we all know that this process is going on today is at the surface of the Earth. The hotter object from which the energy on which life as we know it depends is the Sun, and in this case 'temporary' can be rather long term, since life has existed on Earth for more than three billion years, and our Sun is likely to keep pouring out energy at much the rate it does today for at least another four billion years.

But life also exists deep below the surface of the ocean, around submarine volcanic vents, and even within the rocks of the 'solid' Earth (actually honeycombed with tiny fissures in which heat-loving bacteria thrive). In both cases, the source of energy and order is the heat emerging from the Earth's interior – energy supplied from the decay of radioactive material deep inside the Earth, which has been there since the planet formed. Some biologists speculate that life may even have originated under these conditions, deep inside a planet (not necessarily the Earth), and only later emerged on to the surface, spreading to other planets when meteoritic impacts scattered bits of debris across space.

This understanding of the basics of life also suggests that life may exist in other places quite unlike the surface of the Earth – for example, in clouds of gas and dust in space, feeding off energy from stars.

men and evolution

What are men for? Even assuming that having two sexes involved in reproduction is a good thing, in evolutionary terms, why has evolution produced so many men? The question bothered Charles Darwin, who realized that although the long gestation period means you need plenty of females to ensure survival of the species, only a few men are 'needed' to impregnate them.

The answer came in the 1930s, when the mathematician R. A. Fisher realized that you have to look at the next generation – the grandchildren of the individuals doing the reproducing – and remember that the brutal truth about evolution by natural selection is that the only criterion of success is how many descendants you have. Imagine a situation where on average the first generation produced three times as many girls as boys. If they paired up monogamously for life, two-thirds of the females would not have partners, so they would leave no offspring at all. And even if, at the other extreme, the individuals were completely promiscuous, on average each male would impregnate three females, and end up with three times more offspring than each female.

A mother who just happened, as a result of a genetic mutation, to produce more boys than girls would be a huge evolutionary success, because those males would all get to reproduce. She would have many grandchildren (who would, of course, carry copies of the genes predisposing them to produce more boys than girls). Crucially, she would have many more grandchildren than her sister who produced three girls for every boy.

The same argument works the other way round, with more boys than girls to start with. The logic is obvious, and Fisher proved mathematically that the only stable situation (now called an evolution-arily stable strategy) is to have roughly equal numbers of each sex, whether those individuals end up in monogamous pairs, or harems, or promiscuous relationships.

Milankovitch model

The fact that great ice sheets once covered parts of the globe that are now pleasant to live in is common knowledge – but why did the ice spread out from the poles, and why did it go away again?

The answers are related to the nature of the Earth's orbit around the Sun. The orbit itself stretches slightly, from more circular to more elliptical and back again, with a rhythm some 100,000 years long. Also, there are two cycles related to the way the Earth is tilted in its orbit, and the way it wobbles like a tilted spinning top. One of these cycles is about 40,000 years long, the other about 23,000 years long. What matters is that between them these cycles change the balance of heat across the seasons, although the total amount of heat received by the whole Earth from the Sun in an entire year is always the same.

Ice ages occur when summers in the Northern Hemisphere are cold. Even though this means that winters are relatively warm, it is always cold enough for snow to fall at high latitudes in winter. The important thing is that summers should be cool enough at high latitudes for snow to persist on the ground right through the year. Ice ages end only when all the cycles work together to make summers unusually warm, melting back the ice. This is known as the Milankovitch model, after a Yugoslav astronomer who worked out the details.

With the present day geography of the globe, Ice Age conditions are actually normal, and the kind of conditions we experience today (so-called interglacials) are the exception. In very round numbers, for several million years the pattern has been one of Ice Ages roughly 100,000 years long separated by interglacials some 10,000 to 15,000 years long. The present interglacial began about 10,000 years ago, when the difference between summer and winter was at its greatest. The pattern is now more than halfway back towards Ice Age conditions, with summers getting cooler and winters getting warmer – or at least, things were moving that way, until human beings threw a spanner in the works by pouring greenhouse gases into the atmosphere. The next Ice Age would be due any millennium now, if we weren't interfering with the weather machine.

nanotechnology

Nanotechnology gets its name because it is technology that operates on a scale of a few nanometres, that is, a few billionths of a metre (a few x 10^{-9}m). Human technology is just beginning to be able to manipulate things on this scale – but nature has been doing it for billions of years.

Atoms are roughly spherical objects, each a few tenths of a nanometre across. They form strong connections (chemical bonds) with other atoms, making molecules. In some arrangements, they join together to make long, springy molecules, like pieces of rubber; in others, they are held tightly in place, and so on. They have all the properties needed to build tiny machines out of them. Natural 'nanomachines' operate on the DNA in the chromosomes in your cells, copying bits of genetic material, carrying the instructions around in the cell, and manufacturing new molecules required by the cell. The analogy with an industrial process is so exact that people talked of 'the machinery of the cell' long before the term nanotechnology was invented. And the analogy goes further – bacteria 'swim', for example, with the aid of a lashing tail that works like a propeller.

Starting with nature's nanomachines, the first large-scale applications of nanotechnology are likely to be in medicine and biology – one great hope is to design a tiny robot (a 'nanobot') that can be injected into a person's bloodstream, and will travel around cleaning up the arteries, removing the fatty deposits that cause heart attacks (of course, you would need millions of nanobots to do the job, but they would manufacture themselves out of chemical raw materials).

Slightly further down the line, visionaries foresee nanobots that will get energy from sunlight and use it to rearrange any suitable raw material (such as grass) into a different form (such as steak), cutting out the animal middleman. And the real visionaries talk of tiny robots that will be able to rearrange atoms to build anything out of suitable raw materials and sunlight, essentially for free.

They tell us it could happen within a human lifetime; I live in hope.

natural selection

Charles Darwin's great idea was not evolution. People had known about evolution (from fossils and so on) long before Darwin came along. What they didn't know was how evolution worked. What Darwin came up with was a mechanism for evolution, in his theory of natural selection. It works like this.

When individual members of a species reproduce, the offspring are similar to their parents, but not identical. There is a small amount of variation from one generation to the next. There are also small variations in individual characteristics from one member of a species to another in the same generation (children are slightly different from their parents and from each other), and there are more offspring than needed in each generation ('needed' in order to produce further generations). There is also competition between the individuals in each generation (competition for food, or a mate); overall, the individuals best suited to the prevailing conditions (best fitted to the environment) do best and leave most offspring in their turn. This might be because they are better at finding food, or better at looking after their offspring, or better at attracting a mate.

Since offspring are similar to their parents, any characteristics that make individuals successful are therefore likely to be passed on into succeeding generations. Occasionally, a new ability, or a new variation on an old theme, will emerge in an individual (this is called a mutation). If the new variation is an asset under the prevailing conditions, the individual(s) with the new attribute will do well and leave more offspring, who will inherit the advantage. If it is detrimental, the mutation will die out.

'Fitness', in the Darwinian sense, doesn't necessarily mean physically fit, like an athlete, but fitting into the environment, like the fit of a piece in a jigsaw puzzle.

Olbers' paradox

Why is the sky dark at night? Indeed, why is it so dark in the daytime? This puzzle worried many philosophers and astronomers up to the middle of the twentieth century. One of them was Heinrich Olbers, after whom it is often referred to as 'Olbers' paradox', who discussed it in the nineteenth century.

This is the puzzle. If the Universe is the same everywhere and at all times (what people thought before Edwin Hubble discovered the cosmological redshift effect), then in any direction you look out into space your line of sight should end up on the surface of a star. So, every point on the sky should shine as brightly as the surface of a star, which means as brightly as the Sun.

There are two solutions to the puzzle. Long ago, when the Universe was about half a million years old, it really was as hot, and bright, everywhere as the surface of the Sun today. We know because when we train our detectors on the dark sky what we 'see' is a weak hiss of radio noise, the cosmic background radiation, left over from that fireball. This radiation is cool and feeble today because the Universe has expanded, so the radiation has had to expand along with it, getting thinner as it does so (an extreme version of the redshift effect). This is like the way gas cools when it expands – why gas feels cold when it emerges from a spray can.

The second reason why the sky is dark is that the Universe is not infinitely old. The Big Bang happened about 13 billion years ago, so the oldest stars are only about 12 billion years old. Even pouring out radiation into space for all that time, they have not had anywhere near enough time to fill it with light. The most distant stars (or galaxies) we can see are about 12 billion light years away, because it takes light a billion years to travel a billion light years. To make the sky blaze with light, we would have to be able to see stars a million times further away, so the Universe would have to be a million times older than it is.

Clear evidence that the Universe really did originate at a definite moment in time can be seen with your own eyes, every time you look at the sky.

origin of the Universe

Where did the Universe come from? When astronomers look out into the Universe, they see that it is made up of galaxies like our Milky Way, each composed of hundreds of billions of stars. These galaxies are seen to be receding from each other, so the Universe is expanding. It turns out that this kind of behaviour is exactly predicted by Albert Einstein's general theory of relativity, which says that empty space itself is stretching, and taking galaxies along for the ride.

The implication of both theory and observations is that the Universe was much smaller long ago, and that if you go back far enough everything we can see in the Universe today originated in a hot, dense fireball – the Big Bang. A refinement on this idea, called inflation, explains how that fireball (smaller than our Sun) was produced out of nothing at all by quantum processes. The heat energy from the Big Bang is still detectable today, as a very faint hiss of microwave radio noise coming from all directions in space – the cosmic microwave background radiation.

Inflation theory is still somewhat speculative. But the Big Bang origin itself is highly respectable and described by sound physics. To put this in perspective, the most extreme conditions we can probe on Earth and understand today are those that exist at the heart of an atom, in its nucleus. Winding the observed expansion of the Universe backwards, it seems that the Universe appeared out of nothing at all about 13 billion years ago. That would correspond to a state of infinite density, which nobody claims to understand. But setting that moment as 'time zero', the entire Universe would have been in the nuclear state just one hundred-thousandth of a second later. That was the Big Bang.

At that time, the Universe had a temperature of 1,000 billion °C, and a density 100,000 billion times the density of water. And thoroughly understood, non-controversial physics can explain everything that happened from that time, 0.0001 seconds after time zero, up to the present day.

Planck scale

Quantum theory tells us that no measurable quantity can be completely smooth and continuous. Everything has a grainy structure, which we would see if we could probe down to small enough scales. But these scales are far beyond the reach of any human probes, which is why the world about us seems to be smooth and continuous. The natural quantum scale on which graininess is important is called the Planck scale, in honour of Max Planck, who introduced quantum ideas into physics at the end of the nineteenth century.

The Planck length is the length scale at which space itself is grainy, the 'quantum of length'. It is 10^{-33} cm, about 10^{-20} times the size of a proton. In other words, it would take one hundred billion billion lines, each 1 Planck length long, to stretch across a proton.

The Planck time is the smallest unit of time that can exist (the shortest tick on the cosmic clock). It is equal to the time it would take light, moving at a speed of a few times 10^{10} cm per second, to cross 1 Planck length. That is, in round terms, 10^{-43} seconds. No smaller unit of time has any meaning.

The Planck mass is the mass of a hypothetical particle which would have a wavelength equal to the Planck length. This is about 10^{-5} grams. It sounds small by everyday standards, but don't be fooled – this mass is 10^{19} times the mass of a proton, and would be contained within a volume only 10^{-60} times that of a proton, giving a density (the Planck density) of 10^{79} times the density of an atomic nucleus. It is possible that the Universe was born in this state, as a Planck particle with an age of one Planck time.

plate tectonics

The idea that the continents move about on the surface of the globe has a long history, but was first presented in a proper scientific form by Alfred Wegener in 1912. Wegener was particularly impressed by the way the continents on either side of the Atlantic (especially Africa and South America) fit together like pieces of a jigsaw puzzle if you take the intervening ocean away. But the theory of continental drift only became respectable in the 1960s (under the name 'plate tectonics') when geologists discovered how the continents move.

The key to plate tectonics is the way the Earth's magnetic field reverses from time to time. This does *not* mean that the Earth topples over in space, but that the dynamo inside the Earth fades away then builds up in the opposite sense. Rocks that are laid down by volcanic activity inherit a magnetism appropriate to the time they are being laid down – and the pattern of magnetism in rocks on the ocean floor of the Atlantic (and elsewhere) is one of a series of stripes with alternating magnetism. The order of the stripes is reversed on the opposite side of a line of volcanic activity in the middle of the ocean. This shows that sea floor has been squeezed out on either side of this 'Mid-Atlantic Ridge', like toothpaste being squeezed out of a tube, over tens of millions of years.

The rate at which the Atlantic is getting wider as a result is about 2 cm per year – and this has been measured directly using laser beams bounced off artificial satellites.

To compensate for the widening of the Atlantic (and other spreading ridges), in some parts of the world (notably the western Pacific, near Japan) ocean crust is being destroyed as it is pushed down deep ocean trenches and melts back into the magma below the solid surface of the Earth. The overall effect is like a series of conveyor belts carrying the continents around the globe.

quagga

Forget *Jurassic Park*. Biologists really are on the brink of bringing back to life a species that was once extinct – but it isn't a dinosaur.

The quagga went extinct when the last individual died in an Amsterdam zoo in 1887. But in the early nineteenth century these horse-like creatures roamed the South African plains in large herds. The front of a quagga was striped like a zebra, and the back was brown like a horse. In the 1980s, analysis of strands of DNA taken from a piece of quagga skin kept in a museum showed that it was so closely related to the plains zebra that the two were subspecies of the same evolutionary line. This raised the possibility of bringing the quagga back to life.

In the movie *Jurassic Park*, extinct species were recovered from strands of fossil DNA by 'amplifying' the genetic material biologically. But the quagga went extinct so recently, and is so closely related to the plains zebra, that biologists believe that all of the genes that went to make a quagga still exist in the living cells of different plains zebras today. So a breeding programme is underway at the South African Museum in Cape Town to bring the quagga back from the dead. Starting with plains zebras that have less striping than usual on their hindquarters, the breeding programme is being used to select individuals in each generation whose overall genetic makeup (genotype) most closely matches the DNA in the preserved quagga skin. If the experiment succeeds, it will be the first time that an extinct species has successfully been restored to life; and they are already talking of doing something similar for the dodo.

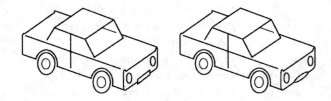

a 'quantum leap' in car design

quantum

One of the scientific terms that has entered popular language is the 'quantum leap'. Curiously, it has almost the opposite meaning in popular language to its scientific meaning. In science, the quantum's most important feature is that it is the smallest possible change that can be made to a system. The other crucial feature of a quantum leap is that if a system, such as an atom, has a choice of states to leap into, it makes its choice about which way to go entirely at random. So a quantum leap is the smallest change it is possible to make, and has been made entirely at random. Something to ponder the next time a politician claims that a new policy represents a 'quantum leap' advance over the old way of doing things.

Red Queen effect

In any ecosystem like that on Earth today, in which many living things interact with one another, as well as with their physical environment, the most powerful driving force for evolution may be – evolution itself. Consider a simple situation in which a fox-like creature feeds exclusively on a kind of rabbit-like creature. Foxes that are able to run faster will be at an advantage, compared with other foxes, because they will catch more rabbits. So genes for running fast will spread in the population of foxes. But rabbits that run faster will be at an advantage, compared with other rabbits, because they will survive longer and have more chance to breed. So genes for running faster will spread in the population of rabbits too. Both species will evolve, but they still end up with just as many foxes eating just as many rabbits.

This is called the Red Queen effect, after the character in Lewis Carroll's *Through the Looking-Glass* who has to keep running as fast as she can in order to stay in the same place (but it doesn't only apply to running, of course; it applies to other biological interactions such as developing sharp teeth to penetrate tough skin, or cunning camouflage as a defence against sharp eyesight).

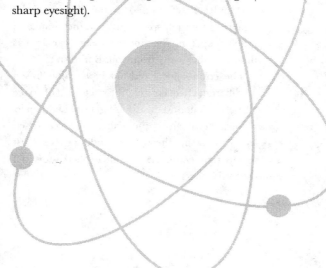

redshift of galaxies

Every smart Alec thinks that he knows what causes the redshift in the light from distant galaxies – and most of them are wrong. Even otherwise respectable textbooks will sometimes tell you that the redshift is a Doppler effect, caused by the motion of the galaxies through space, just like the change in pitch deepening the tone of a police car siren as the car rushes away from you. But it isn't.

The cosmological redshift is actually caused by a different effect entirely, the stretching of space itself between us and the distant galaxy. Light on its way towards us from the galaxy gets stretched because it is passing through expanding space – just as a picture painted on a rubber sheet gets distorted if you stretch the sheet.

This is what tells us that the Universe is expanding, and must have been much smaller long ago – key evidence that it was born in a Big Bang. But there is another peculiarity about the cosmological redshift.

The redshift gets its name because the stretching makes wavelengths of light longer, and red light has longer wavelengths than blue light. But it happens that some of the most distant objects in the Universe, the quasars, radiate most of their energy as very short-wavelength ultraviolet light. After the redshift has done its work, this energy is shifted into the blue part of the spectrum, because blue light has longer wavelengths than ultraviolet light. So the effect of the redshift on many quasars is to make them look blue.

second law
of thermodynamics

> The second law of thermodynamics holds, I think, the supreme position among the laws of Nature. If someone points out to you that your pet theory of the universe is in disagreement with Maxwell's equations – then so much the worse for Maxwell's equations. If it is found to be contradicted by observation – well, these experimentalists do bungle things sometimes. But if your theory is found to be against the second law of thermodynamics I can give you no hope; there is nothing for it but to collapse in deepest humiliation.
>
> **Arthur Eddington**,
> *The Nature of the Physical World*, 1935

Why is the second law so pre-eminent (and why isn't it the first law)? It isn't the first law because that law is a kind of throat-clearing statement that heat is a form of energy. It is pre-eminent because it says that things wear out. One way of expressing this is that heat only flows from a hotter object to a cooler object, unless energy is put into a system from outside. Ice cubes in a glass of water melt; we make ice cubes in a fridge by providing energy (so that if a fridge is on full blast with the door open in a sealed room, the room gets hotter, not colder, because of the energy used by the motor of the fridge).

Another expression of the second law is that the amount of disorder (also known as entropy) in the Universe increases. The only way to make a pocket of order in a disorderly world is the way life on Earth does – by feeding off an outside source of energy (in our case, the Sun).

The second law defines the arrow of time. The future is the direction of time in which the ice cubes melt, buildings fall down, and cars rust. The future is also the direction of time away from the Big Bang in which the Universe was born. There may be a deep truth concealed in this 'coincidence', but nobody has yet been able to work out what it is.

string

Particle theorists are currently excited about string, not particles. They tell us that the world is not, after all, made up of tiny point-like entities, or miniature billiard balls, but one-dimensional entities which make loops on the Planck scale, 10^{-33} cm across, some 10^{20} times smaller than an atomic nucleus. The scale on which these strings operate is as much smaller than an atom as an atom is smaller than the Solar System.

The key feature of these strings is that they are not regarded as one-dimensional in the everyday sense of the term. Instead, they involve oscillations in more dimensions than the three of space and one of time that we are used to. These 'extra' dimensions, although essential in determining the properties of the strings, are thought of as hidden away from sight by compactification.

The usual example of compactification is to think of a hosepipe viewed from a distance. It looks like a one-dimensional line. But look closer and you see that it is made of a two-dimensional sheet (which might ripple in interesting ways), wrapped around the third dimension. The same trick can be used mathematically to wrap up any number of dimensions you like, shrinking them down and leaving only the familiar four dimensions visible.

In string theory, there are no different particles, just these tiny loops of string (involving ten or more dimensions) vibrating in different ways. Just as you can play different notes on a single violin string by making it vibrate in different ways, one kind of string vibration might correspond to an electron, one to an up quark, one to a photon, and so on.

Then, in the mid 1990s, along came membranes (often referred to as M-theory). What if the strings were actually rolled up out of a sheet, or membrane? The extra dimension is immediately rolled up out of the way so that the membrane behaves like the ten dimensional string of the once-standard string theory. The idea is taken seriously because M-theory offers a unique package to describe all of the forces and particles of nature. There were previously several different versions of string theory, but it turns out that they are all manifestations of a single M-theory. A large reward for the small price of adding just one more dimension to the story.

tachyons

Although no particle can be accelerated from below the speed of light to faster than the speed of light, the equations of the special theory of relativity tell us that there could be particles which always travel faster than light. Such 'superluminal' particles have never been detected, but they have been given a name – 'tachyon'.

The theoretical behaviour of tachyons mirrors the behaviour of ordinary 'subluminal' particles, so that, for example, whereas a subluminal particle can never be accelerated up to the speed of light, tachyons can never slow down below the speed of light. An ordinary particle moving at high speed will tend to lose energy and slow down, but a tachyon will tend to lose energy and speed up – it is as if both subluminal and superluminal particles are repelled from the speed of light. If a tachyon were created in some violent event in space, it would radiate energy away furiously (perhaps in the form of gravitational radiation, or as electromagnetic radiation) and go faster and faster until it had zero energy (and was therefore undetectable) and was travelling at infinite speed.

Most physicists believe that tachyons do not exist, and that the solutions of the equations which describe them are as meaningless as the 'negative roots' that come out of simple quadratic equations. For example, it is possible (if rather silly) to describe the number of people taking part in a game of bridge as being 'the square root of 16'. The square root of 16 may be either +4 or -4, but it is obvious from the context that there are not '-4' people seated at the card table.

If tachyons did exist, however, they would have one other curious property – they would travel backwards in time. This has encouraged some physicists studying cosmic rays to look for 'precursor' events, in which a particle from space arrives just before a cosmic-ray burst. The argument is that if the cosmic-ray burst originates when a high-energy particle from space strikes the top of the Earth's atmosphere, then any tachyons created at the same time will reach the ground just before the cosmic ray burst is created, having travelled backwards in time during their journey down to the ground. Unfortunately, no conclusive evidence for tachyonic precursors of this kind has been found.

theory

The most misunderstood term in science is the word 'theory'. Cranks who find it impossible to believe in ideas such as the special theory of relativity or evolution by natural selection sometimes point out that 'even the scientists say that it is "only a theory"'. They refer to the way we use the word in everyday language, to mean a (possibly crackpot) speculation – my brother has a theory that the England football team would score more goals if they wore longer shorts. But in science, the name for such an unproven, possibly crackpot, idea is a hypothesis. A theory is an idea that was once a hypothesis, but has been tested by experiment and by observation of the real world, *and has passed every test that it has ever been subjected to*.

As soon as a theory fails an experimental or observational test, strictly speaking it has to be replaced by a better, more complete theory. But the old theory may still be useful in a restricted area, provided you know its limitations. The best example is gravity. Isaac Newton's theory of gravity can be used to explain things like the orbit of the Moon around the Earth, or predict the trajectory of a baseball hit by a batter. It can't explain properly what goes on in very strong gravitational fields, or how light is bent by the Sun. Albert Einstein's general theory of relativity explains all the things Newton's theory does, and also explains what happens with strong gravitational fields, and how gravity bends light. So it is a better theory. But it is still easier to use Newton's theory if you just want to work out the trajectory of a baseball.

unicellular life

People like to think that they are the most successful form of life on Earth, the result of billions of years of evolution by natural selection, forever improving the individuals that it operates on. This is wrong. Evolution is not about improving anything, it is about producing many offspring that carry copies of your genes forward into later generations. By this criterion, the most successful forms of life on Earth are the unicellular life-forms, micro-organisms which have been around for billions of years without ever having to change their basic form.

At every stage of evolution, successful species become superbly adapted to their ecological niches, and stay much the same. We are descended from a long line of evolutionary misfits and outcasts, the individuals who were less successful in those particular niches, and had to find new ones to occupy. Take the transition from the sea to the land. The most successful fish stayed as fish – there was no pressure for them to move on to the land.

It was the less successful fish who were forced into the shallows and which found a new way of life by becoming amphibians. Then, the less successful amphibians got pushed out of the water altogether, and had to become reptiles; and so on.

All the while, simple bacteria were doing what they had always done – surviving, reproducing and filling all the available niches in abundance. Close relatives of the earliest bacteria, single-celled organisms that lack even a central nucleus in which to package the DNA of their cells, are the most abundant and widespread organisms on Earth. Because of the way unicellular organisms reproduce, by simple fission producing two daughter cells which are each identical to the mother, it isn't even really true to say that the cells around today are the descendants of the original bacteria. In a sense, some of the individual bacteria alive on Earth today are the same individual cells that have been here for several billion years.

Universal distance scale

Galaxies are much closer together than stars are, if you look at them in the appropriate way. Imagine the Sun being reduced to the size of an aspirin. On that scale, the nearest star would be another aspirin 140 km away. The distance from one star to another is tens of millions of times the diameter of an individual star, and our Milky Way Galaxy is made up of several hundred billion stars, orbiting around a common centre, separated by these vast distances but held in orbit by gravity.

Now move out from stars to galaxies. Imagine the whole Milky Way Galaxy, several hundred billion stars in a disk, being shrunk to the size of an aspirin. On that scale, the nearest comparable large galaxy to us, the Andromeda Galaxy, would be another aspirin just 13 cm away. This is a bit misleading, because both the Andromeda Galaxy and the Milky Way are part of a small system of galaxies known as the Local Group – it's a bit like saying the Moon is the nearest planet to Earth. But even the distance to the nearest comparable group of galaxies would only be 60 cm, on the aspirin scale. Just 3 metres away, on the aspirin scale, we would see a swarm of more than 2,500 galaxies, spread over a volume about the same size as a basketball – the Virgo Cluster of galaxies. Indeed, the whole visible Universe, containing several hundred billion galaxies, would be confined in a sphere just 1 km across, on this scale

If galaxies were as far apart, relatively speaking, as stars are, we would not yet have seen one per cent of the distance out to the nearest galaxy, and we would not know that they existed. The Universe really is a crowded place, packed full of galaxies like the Milky Way.

virtual particles

There is more to empty space than you might imagine. Einstein's famous equation $E = mc^2$ tells us, among other things, that particles can be manufactured out of pure energy. An energetic gamma ray, for example, can turn into a pair of particles – an electron and a positron (it has to be a particle and its antiparticle counterpart, to balance the quantum equations).

Werner Heisenberg's uncertainty principle is almost as famous, but fewer people have much idea what it means. It tells us that nothing is precisely known in the quantum world (not just that it is impossible to measure things like position and velocity precisely, but that entities like electrons literally do not have precise positions and velocities). Among other things, the uncertainty relation tells us that the Universe, or 'Mother Nature', or whatever you want to call it, does not know how much energy there is associated with a small volume of space for a small time.

This means that even in a vacuum with zero energy, particle–antiparticle pairs (such as electron–positron pairs) can pop into existence out of nothing at all, provided they almost immediately annihilate one another (in a sense, before the Universe notices they are there), and give back the energy they have borrowed from the vacuum. Such short-lived quantum entities are called virtual particles.

There is an adage in quantum physics that 'anything which is not forbidden is compulsory'. Because virtual particles are allowed to exist, the quantum vacuum actually seethes with these particles. Near a negatively charged real particle (such as an electron), virtual electrons are, even during their short lives, slightly pushed away by the electron, and virtual positrons are slightly pulled towards it. This 'polarization of the vacuum' affects the measured properties of an electron (in particular, its measured electric charge). The size of the effect is predicted by quantum theory, and it exactly matches experimental results. This is one of the triumphs of quantum physics, and shows that virtual particles are real. Physicists can measure the influence of nothing on electrons.

viruses

Viruses are the nearest thing in nature to zombies – the living dead. A virus consists of nothing more than a small quantity of genetic material wrapped in a coat of protein. Crucially, it cannot reproduce (make copies of itself) unaided, or even with the aid of another virus. So it fails a fundamental test of life. But a virus, which is much smaller than a cell, can attach itself to a cell and make a hole in the wall of the cell, through which the virus injects its genetic material. The cell that is being attacked might be one of the many cells in a plant or an animal, or it might be a bacterium (viruses that attack bacteria are called bacteriophages, or phage for short). When viral genetic material gets into the cell, the machinery of the cell is fooled into using the viral genetic material to make copies of the original virus.

Sometimes, all that happens is that bits of virus genetic material get integrated into the cell's genetic material, so that copies of the virus are made and released without doing much damage to the working of the cell. In other cases, the cell's normal activity completely shuts down, and it simply turns out viruses until all the chemical raw materials it has are used up. The

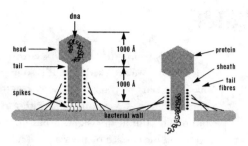

a phage at work
one Ångstrom is one hundred-millionth (10^{-8}) of a centimetre

viruses may escape from the cell through a small hole, or (usually in the case of bacteria attacked by phage), the cell bursts and releases the viruses in a flood. In some circumstances, viruses produced in this way may carry bits of the cell's own DNA with them, and can even 'infect' other cells with this, so that genes can be transferred by viruses from one cell to another (even, under rare circumstances, from one species to another).

Using this ability of viruses, researchers have managed to use a 'deactivated' virus to introduce a healthy version of the gene that, when faulty, causes cystic fibrosis into the lungs of rats. They hope that it may soon be possible for human sufferers from cystic fibrosis to inhale a spray every few weeks that will keep the cells in their lungs functioning normally.

water

Water is weird stuff. It only seems ordinary to us because we are used to it. It is so important to life that our very term for a region devoid of water – desert – is synonymous with a region devoid of life. But at first sight, it has no right to be a liquid on the surface of the Earth at all.

By and large, the lighter a molecule is, the colder it must be before it will condense from the gas state to a liquid. Water is a pretty light molecule, weighing just 18 units on a scale where a hydrogen atom weighs 1 unit. But there are much heavier gases (such as carbon dioxide, molecular weight 44) which are gases under the same conditions where water forms oceans of liquid.

The reason is that water is very good at forming a kind of link called a hydrogen bond. It happens because hydrogen atoms each have only one negatively charged electron shielding their positively charged nucleus. When this electron links up with oxygen in water molecules as H_2O, the negative charge lies between the hydrogen nucleus and the oxygen atom, leaving bare positive charge exposed. This doesn't happen for any other atoms, because they all have additional electrons to shield their nuclei.

The excess positive charge from hydrogen at one end of a water molecule is attracted to negative electrons at the oxygen end of other water molecules. This makes water molecules sticky, so that they cling on to one another instead of bouncing off easily. And that is why water is a liquid at room temperature and pressure, while carbon dioxide is not.

Even stranger, when water freezes, the shape of its molecules is just right for hydrogen bonds to hold the molecules in a very open lattice, rather like the structure of diamond but not so strong. The lattice is so open that the density of solid water (ice) is less than that of liquid water. So ice floats. If it sank – like any self-respecting solid in its own liquid (for example, iron) – in winter ice would form at the bottom of a pond or ocean, not at the top. Instead of having ice as an insulating lid holding the remaining heat in, oceans would lose all their heat and freeze solid, from the bottom up – and there would be no life on Earth.

wormhole

How does the Universe make the laws of physics the same everywhere? We know that the laws are the same everywhere, by studying the light from distant objects in different directions in space. The nature of the spectrum of light shows that the atoms which make up quasars and galaxies on one side of the sky are behaving in exactly the same way as the atoms that make up quasars and galaxies on the opposite side of the sky. But no information can travel through space faster than the speed of light. If something happened to change the way atoms work on one side of the sky, how quickly could the rest of the Universe respond?

The boring answer is that there is probably no need to worry about this, because the laws of physics were specified (somehow!) in the Big Bang, and cannot change. But a respectable speculation suggests that there may be another layer to this puzzle. Quantum physics says that, just like everything else, space itself (or, strictly speaking, space-time) cannot be continuous, but must have a grainy structure on the smallest scales. In this case, the appropriate scale is the

Planck length, about 10^{-33} cm. The idea is that on this scale the structure of spacetime is made up of countless tiny 'wormholes' – tunnels through space and time with ends like miniature black holes. These wormholes would form a tangled mess, like cosmic spaghetti, which forms the structure of what looks to us a smooth and continuous time and space.

In this picture, the mouth of such a wormhole in our part of the Universe could be linked with a mouth anywhere else in the Universe. It is a curious feature of wormholes that, according to the general theory of relativity, it takes no time at all for anything to travel from one end to the other, no matter where the two ends are located. The mouths are vastly smaller than an atom, or even a subatomic particle such as a proton, so nothing material could get through them. But maybe information could leak through the wormholes from one part of the Universe to another, providing a means for different regions of spacetime to keep in touch with one another, and ensuring that the laws of physics in all parts of the Universe march to the beat of the same drum.

X-factor

You might think that the prospects for the human race are now largely within our control – even things like wars and the risk of global warming are our own fault, and we ought to be able to do something about them. Optimists even argue that we have the technology to stop an ice age in its tracks, by spreading soot on the ice sheets to make them dark, so they absorb heat from the Sun and melt. But there is an X-factor – the risk of disaster striking without warning from the skies. Your risk of being killed like this is equal to the risk of being killed in an airline accident. But if you go, so will a lot of other people.

An impact big enough to cause a global crop failure happens once every 300,000 years, and you have a 1 in 4 chance of dying if such an impact occurs in your lifetime. So the annual risk of death from this cause is 1 in $(4 \times 300,000)$, which is 1 in 1.2 million. If the average human lifetime is 75 years, the overall chance of being killed in this way is 75 times greater – 75 in 1.2 million, or 1 in 20,000.

Statistically, the chance of any one person being killed by a major comet impact in any one year may be only about 1 in

2 million, but that still means that on average 2,700 people are killed by large comets each year, 390 of them in the developed world alone. You just have to take the average over a long enough time.

The cost of road-safety measures in Britain in the mid 1990s worked out at just over £800,000 for each life that was saved. If we cared as much about the comet and asteroid hazard as we did about road safety, that would imply that the developed world ought to be spending nearly $500 million a year (enough to make four movies on the scale of *Titanic*) to prevent cosmic impacts. But hardly anything is actually being spent on the problem. The fact that nobody is killed by such an impact for many years, and then many people are killed at once, does not alter the argument. There are long stretches of road where nobody is killed for many years, and no big accidents occur; but that is not a good argument for saving costs by not bothering to maintain safety barriers and decent lighting on those stretches of road.

X-particle

The favoured theories of how the forces and particles of nature work say that the first particles that were produced in the Big Bang, in which the Universe was born, each had a mass one million billion times the mass of a proton. They are known as X-particles. Within a billionth of a billionth of a billionth (10^{-27}) of a second, the X-particles decayed, fragmenting into showers of other particles and antiparticles.

In experiments at the kinds of energies available on Earth (or anywhere in the Universe) today, particles can be made out of pure energy or by the decay of other particles, but only if particles and so-called antiparticles are created in equal numbers. When an antiparticle meets its particle counterpart, the pair annihilate one another in a burst of electro-magnetic energy.

If X-particles had behaved like that, soon after the Big Bang all the particles and antiparticles would have annihilated one another. There would have been no matter in the Universe, only light. But the theory says that there is a tiny unevenness in the laws of physics, which allowed X-particles to decay in such a way that they produced a billion and one particles for every billion antiparticles. So after the annihilation had finished, the one in a billion particles were left over to form the stars, galaxies, planets and people in the Universe today.

Everything else annihilated – to produce a billion photons for every material particle in the Universe. Although it is hard to measure (or estimate) such things precisely, the Universe is indeed filled with a sea of radiation left over from the Big Bang (the cosmic microwave background radiation) and there do seem to be about a billion photons in the Universe for every particle estimated to be in all the bright stars and galaxies. This makes X-particle theory respectable, although still speculative – and implies that you and I are part of a one in a billion bit of leftover stuff from the moment of creation.

Y chromosome

The Y chromosome is the odd one out in the human genome. All the genes that describe the construction and maintenance of the human body are contained in chromosomes, which are little rod-like packages of DNA. There are 23 pairs of chromosomes in every cell of the human body (except the special cells involved in reproduction), one member of each pair copied from an original inherited from one parent, the other carrying copies of genes from the other parent. In all but one of the pairs, the chromosomes are structurally the same as each other, although the gene for eye colour, say, on one chromosome may be different from the one on its partner. In women, this similarity extends to the sex chromosomes, both of which have a cross shape, like a fat letter X. They are called X chromosomes. But in men's cells, a single X chromosome is paired up with a kind of stunted version, with one arm of the X missing; this is called a Y chromosome. Except in people with genetic disorders, Y chromosomes are never paired with each other.

It is the presence of the Y chromosome in the cells of men that makes them men. Without a Y chromosome, a developing foetus develops ovaries and becomes female; with the Y chromosome present, the same organs develop along a different path to become testes, and the body becomes male. Among other things, this means that physical sexuality is inherited solely from your father. Your mother only has X chromosomes to pass on, as far as sex is concerned, so you have to inherit one from her. But your father has an X and a Y, so there is a 50:50 chance of getting either, an even-handed lottery of life that decides whether a baby is a boy or a girl.

This isn't quite all that the Y chromosome does, though. The X chromosome carries several genes in addition to helping determine sex; but the Y chromosome (only one-fifth as big as the X), carries just one – a gene that determines whether or not the body it inhabits has hairy ears.

Young's double-slit experiment

The central mystery of the quantum world is encapsulated in the modern version of an experiment devised by Thomas Young in the early nineteenth century. Young's double-slit experiment (also known as the experiment with two holes), in its original form, involved shining light through two slits in a screen (razor-blade cuts in a sheet of cardboard are ideal) and on to a white surface. The light makes a stripy pattern of light and shade which is interpreted as the result of waves spreading out from each of the two holes and interfering with one another, like ripples interfering on a pond. It proves that light travels as a wave.

Unfortunately, in the early twentieth century, other experiments showed light behaving as a stream of particles (photons). The mystery deepened when it was discovered that electrons (previously regarded as particles) also behave like waves, and make interference patterns.

In the latest versions of the experiment with two holes, individual photons or individual electrons are fired, one at a time, towards the two holes. (I stress that these experiments really have been carried out, exactly as I describe.) The quantum entities are detected on the other side by a device like a TV screen, building up an image. In each case, the quantum entities start out as particles. In each case, each one arrives at the detector screen as a particle and makes a single spot on it. But as hundreds and then thousands of 'particles' are fired through the experiment, the spots on the detector screen build up to make the interference pattern that belongs to waves going through both holes at once. Not only does each particle seem to pass through both holes at once and interfere with itself, but somehow it knows where it must go in the overall pattern to build up this image, as if each 'particle' is influenced by past and future events.

Don't worry that this doesn't match up with everyday common sense. Nobody knows how the quantum world can be like that. But it is.

zero energy Universe

The one equation everybody knows is $E = mc^2$. Mass is equivalent to energy. There are hundreds of billions of stars like our Sun in a galaxy like our Milky Way, and hundreds of billions of galaxies in the visible Universe. So the total energy of the Universe must be absolutely enormous, right? Wrong – the total energy of the Universe is zero.

Imagine breaking any material object (a star, your wrist-watch, a pint of lager) into its fundamental component pieces. Now spread the pieces out as far as you can, so they are infinitely far apart. The gravitational force they exert on each other is proportional to 1 divided by the square of the distance between them. Anything divided by infinity is zero; 1 divided by infinity^2 is definitely zero. But now imagine giving the pieces a little nudge, so they start falling together under the influence of gravity. When things fall like this, they gain energy, which makes them move faster. This is how stars get hot – clouds of gas in space collapse under gravity and heat up because the atoms they are made of move faster. The energy comes from the gravitational field of the material. But you started with zero gravitational energy, and you have taken some away. So now, the energy of the gravitational field is less than zero – it is negative.

Carry the calculation through using the general theory of relativity, and you find that when the material has fallen together at a point the amount of energy in the gravitational field is precisely *minus mc*². It exactly balances the Einsteinian mass energy involved!

This is how the Universe could have appeared at a point, out of nothing at all – because it contains no energy at all. The Universe only looks so interesting because its zero overall energy is divided evenly between a lot of mass and a lot of negative gravitational energy.

Zweig

George Zweig ought to be one of the most famous names in science. He isn't because he made the mistake of having a brilliant idea while he was still a student, and nobody took him seriously. His brilliant idea, in 1963, was that the so-called 'fundamental' atomic particles – the proton and the neutron – might each be made up of three truly fundamental entities, which he called 'aces'. The idea was laughed out of court by his superiors, one of whom described it as 'the work of a charlatan'. Around the same time, Murray Gell-Mann, a senior and respected physicist, came up with the same idea, and called the particles 'quarks'. Even though Zweig's version is simpler and easier to understand, a few years later when evidence that protons and neutrons are not fundamental was found from experiments, it was Gell-Mann's name that was given to the particles within them. The moral? Get to be a respected professor first, before presenting unconventional ideas to the world of science.

index

107

THE LITTLE BOOK OF SCIENCE

Dr John Gribbin trained as an astrophysicist at the University of Cambridge before becoming a full-time science writer. He has worked for the science journal Nature, and the magazine New Scientist and has contributed articles on science topics to *The Times*, the *Guardian* and the *Independent*, and has made several acclaimed science series for BBC Radio 4. John Gribbin has received awards for his writing in both Britain and the United States and is currently a visiting Fellow in astronomy at the University of Sussex. His many books include *In Search of Schrödinger's Cat, Stephen Hawking: A Life in Science* (with Michael White) and *In Search of SUSY*. John Gribbin is also the author of several science fiction works including *Innervisions*.

He is married with two sons and lives in East Sussex.

Frommer's®

Italian
PhraseFinder &
Dictionary

2nd Edition

WILEY

Wiley Publishing, Inc.

Published by:

Wiley Publishing, Inc.

111 River St.
Hoboken, NJ 07030-5774

ISBN-13: 978-0-470-93649-8

Italian Editor: Donald Strachan
Series Editor: Jessica Langan-Peck
Photo Editor: Richard H. Fox
Cover design by Paul Dinovo

Translation, Copyediting, Proofreading, Production, and Layout by:
Lingo Systems, 15115 SW Sequoia Pkwy, Ste. 200
Portland, OR 97224

For information on our other products and services or to obtain technical support, please
contact our Customer Care Department within the U.S. at 800/762-2974, outside the U.S. at
317/572-3993 or fax 317/572-4002.
Wiley also publishes its books in a variety of electronic formats. Some content that appears
in print may not be available in electronic formats.

Manufactured in the United States of America

5   4   3   2   1

Contents

An Invitation to the Reader

In researching this book, we discovered many wonderful sayings and terms useful to travelers in Italy. We're sure you'll find others. Please tell us about them, so we can share the information with your fellow travelers in upcoming editions. If you were disappointed with an aspect of this book, we'd like to know that, too. Please write to:

Frommer's Italian PhraseFinder & Dictionary, 2nd Edition
Wiley Publishing, Inc.
111 River St. • Hoboken, NJ 07030-5774

An Additional Note

The packager, editors, and publisher cannot be held responsible for the experiences of readers while traveling. Your safety is important to us, however, so we encourage you to stay alert and be aware of your surroundings. Keep a close eye on cameras, purses, and wallets, all favorite targets of thieves and pickpockets.

Frommers.com

Now that you have the language for a great trip, visit our website at **www.frommers.com** for travel information on more than 3,000 destinations. With features updated regularly, we give you instant access to the most current trip-planning information available. At Frommers.com, you'll also find the best prices on airfares, accommodations, and car rentals—and you can even book travel online through our travel booking partners. At Frommers.com, you'll also find:

* Online updates to our most popular guidebooks
* Vacation sweepstakes and contest giveaways
* Newsletter highlighting the hottest travel trends
* Online travel message boards with featured travel discussions

INTRODUCTION: HOW TO USE THIS BOOK

More than 70 million people speak Italian, a language so melodious it can make a chore list resemble an opera aria. This rich tongue is already somewhat familiar to most chefs, artists, musicians, and others who savor Italy's rich cultural heritage.

As a Romance language, Italian is closely related to Latin, French, Italian, Portuguese, and Romanian. The modern nation of Italy is a patchwork of former city-states—and its dialects still reflect these strong regional loyalties. Thus you may hear variations on the phrases found here. But rest assured: Native speakers have carefully reviewed all the material.

Our intention is not to teach you Italian: A class or audio program is best for that. Rather, we offer an easy-to-use travel tool. You don't need to memorize the contents or flip frantically to locate a topic, as with other books. Rather, Frommer's has fingertip referencing and an extensive PhraseFinder dictionary at the back.

Say a taxi driver accidentally hands you $5 instead of $10. Look up "change" in the dictionary and discover how to say: "Sorry, but this isn't the right change." Then follow the cross-reference to numbers, so you can explain exactly how much is missing.

This book may even be useful to advanced students, because it supplies speedy access to exact idioms. Elegance and **bella figura**—cutting a fine figure—are important to Italians. So is the right phrase at the right time.

As Tim Parks observed in **An Englishman in Verona**: "While Italians usually seem to like foreigners, the foreigners they like most are the ones who know the score, the ones who have caved in and agreed that the Italian way of doing things is best. . . . There is an order to follow in all things; follow it, even when it borders on the superstitious and ritualistic."

Thus, from the novice to the conversationally adept, travelers can benefit from the detailed conversation and etiquette tips here. At each turn, we've researched the traditions and travails, striving to make this phrasebook accessible, practical, and indispensable.

We've also tried to make this volume useful to all English speakers, though Frommer's is based in the United States. As with all our publications, we hope this enriches your travel experience greatly. And with that, we offer this wish: **Fate il miglior viaggio possibile!**

CHAPTER ONE

If you tire of toting around this phrasebook, tear out or photocopy this chapter. You should be able to get around using only the terms found in the next 35 pages.

BASIC GREETINGS

For a full list of greetings, see p111.

Hello.	**Salve.**
	SAHL-veh
How are you?	**Come sta?**
	KOH-meh stah
I'm fine, thanks.	**Bene, grazie.**
	BEH-neh GRAH-tsyeh
And you?	**E lei?**
	eh lay
My name is ____.	**Mi chiamo ____.**
	mee KYAH-moh
And yours?	**E lei?**
	eh lay
It's a pleasure to meet you.	**Piacere di conoscerla.**
	pyah-CHEH-reh dee KOH-noh-shehr-lah
Please.	**Per favore.**
	pehr fah-VOH-reh
Thank you.	**Grazie.**
	GRAH-tsyeh
Yes.	**Sì.**
	SEE
No.	**No.**
	noh

1

OK.

No problem.

I'm sorry, I don't understand.

Would you speak slower, please?

Would you speak louder, please?

Do you speak English?

Do you speak any other languages?

I speak _____ better than Italian.

Would you spell that?

Would you please repeat that?

Would you point that out in this dictionary?

OK. / Va bene.
OK / vah BEH-neh

Nessun problema.
nehs-SOON proh-BLEH-mah

Mi dispiace, non capisco.
mee dee-SPYAH-cheh nohn kah-PEE-skoh

Può parlare più lentamente, per favore?
PWOH pahr-LAH-reh PYOO lehn-tah-MEHN-teh pehr fah-VOH-reh

Può parlare a voce più alta, per favore?
PWOH pahr-LAH-reh ah VOH-cheh PYOO AHL-tah pehr fah-VOH-reh

Parla inglese?
PAHR-lah een-GLEH-seh

Parla altre lingue?
PAHR-lah AHL-treh LEEN-gweh

Parlo _____ meglio dell'italiano.
PAHR-loh _____ MEH-lyoh dehl-lee-tah-LYAH-noh

Può dirmi come si scrive?
PWOH DEER-mee KOH-meh see SKREE-veh

Può ripetere, per favore?
PWOH ree-PEH-teh-reh pehr fah-VOH-reh

Può indicarlo su questo dizionario?
PWOH een-dee-KAHR-loh soo KWEHS-toh dee-tsyoh-NAH-ryoh

THE KEY QUESTIONS

With the right hand gestures, you can get a lot of mileage from the following list of single-word questions and answers.

Who?	**Chi?**
	kee
What?	**Cosa?**
	KOH-sah
When?	**Quando?**
	KWAHN-doh
Where?	**Dove?**
	DOH-veh
Why?	**Perché?**
	pehr-KEH
How?	**Come?**
	KOH-meh
Which?	**Quale?**
	KWAH-leh
How many / much?	**Quanto -a? / Quanti -e?**
	KWAHN-toh -tah / KWAHN-tee -teh

THE ANSWERS: WHO

For full coverage of pronouns, see p19.

I	**Io**
	EE-oh
you	**Lei / tu**
	lay / too
him	**lui**
	LOO-ee
her	**lei**
	lay
us	**noi**
	NOH-ee
them	**loro**
	LOH-roh

THE ANSWERS: WHEN

For full coverage of time, see p12.

now	**ora**
	OH-rah
later	**più tardi**
	PYOO TAHR-dee
afterwards	**dopo**
	doh-poh
earlier	**prima**
	pree-mah
in a minute	**fra un minuto**
	frah oon mee-NOO-toh
today	**oggi**
	OHD-jee
tomorrow	**domani**
	doh-MAH-nee
yesterday	**ieri**
	YEH-ree
in a week	**fra una settimana**
	frah oonah seht-tee-MAH-nah
next week	**la settimana prossima**
	lah set-tee-MAH-nah prohs-SEE-mah
last week	**la settimana scorsa**
	lah set-tee-MAH-nah SKOHR-sah
next month	**il mese prossimo**
	eel MEH-zeh prohs-SEE-moh
At _____	**Alle _____**
	AHL-leh
ten o'clock this morning.	**dieci di stamattina.**
	DYEH-chee dee stah-maht-EE-nah
two o'clock this afternoon.	**due di oggi pomeriggio.**
	DOO-eh dee OHD-jee poh-meh-REED-joh
seven o'clock this evening.	**sette di stasera.**
	SEHT-teh dee stah-SEH-rah

For full coverage of numbers, see p7.

THE ANSWERS: WHERE

here	**qui / qua** *kwee / kwah*
there	**lì / là** *LEE / LAH*
near	**vicino** *vee-CHEE-noh*
closer	**più vicino** *PYOO vee-CHEE-noh*
closest	**il più vicino** *eel PYOO vee-CHEE-noh*
far	**lontano** *lohn-TAH-noh*
farther	**più lontano** *PYOO lohn-TAH-noh*
farthest	**il più lontano** *eel PYOO lohn-TAH-noh*
across from	**di fronte a** *dee FROHN-teh ah*
next to	**di fianco a** *dee FYAHN-koh ah*
behind	**dietro a** *DYEH-troh ah*
straight ahead	**diritto** *dee-REET-toh*
left	**a sinistra** *ah see-NEES-trah*
right	**a destra** *ah DEHS-trah*
up	**su** *soo*
down	**giù** *JOO*
lower	**più giù** *PYOO JOO*
higher	**più su** *PYOO soo*
forward	**avanti** *ah-VAHN-tee*
back	**indietro** *een-DYEH-troh*

around	**attorno** *aht-TOHR-noh*
across the street	**dall'altra parte della strada** *dahl-LAHL-trah PAHR-teh DEHL-la STRAH-dah*
down the street	**più avanti** *PYOO ah-VAHN-tee*
on the corner	**all'angolo** *ahl-LAHN-goh-loh*
kitty-corner	**all'angolo opposto** *ahl-LAHN-goh-loh ohp-POHS-toh*
_____ blocks from here	**_____ traverse più in là** *_____ trah-VEHR-seh PYOO een LAH*

For a full list of numbers, see the next page.

THE ANSWERS: WHICH

this one	**questo -a** *KWEHS-toh -tah*
that	**quello -la** *KWEHL-loh -lah*
these	**questi -e** *KWEHS-tee -teh*
those	**quelli -le** *KWEHL-lee -leh*

HELP/EMERGENCIES

Can you help me?	**Mi può aiutare?** *Mee poo-oh ayu-TAH-reh?*
Help!	**Aiuto!** *AJU-toh*
Call the police!	**Chiami la polizia** *KEE-ah-mee lah poh-lee-tzeeah*
I need a doctor.	**Ho bisogno di un dottore.** *Oh bee-SOHN-nyoh dee oon doht-TOH-reh*
Thief!	**Al ladro!** *Ahl LAH-droh*
My child is missing.	**Mio figlio si è perso.** *Meeh-oh FEEL-lyoh see eh PEHR-soh*
Call an ambulance.	**Chiami un'ambulanza.** *KYAH-mee OO-nahm-boo-LAHN-tzah*

NUMBERS & COUNTING

one	**uno** / *OO-noh*	seventeen	**diciassette** / *dee-chahs-SEHT-teh*
two	**due** / *DOO-eh*	eighteen	**diciotto** / *dee-CHOT-toh*
three	**tre** / *treh*	nineteen	**diciannove** / *dee-chahn-NOH-veh*
four	**quattro** / *KWAHT-troh*	twenty	**venti** / *VEHN-tee*
five	**cinque** / *CHEEN-kweh*	twenty-one	**ventuno** / *vehn-TOO-noh*
six	**sei** / *SEH-ee*	thirty	**trenta** / *TREHN-tah*
seven	**sette** / *SEHT-teh*	forty	**quaranta** / *kwah-RAHN-tah*
eight	**otto** / *OHT-toh*	fifty	**cinquanta** / *cheen-KWAN-tah*
nine	**nove** / *NOH-veh*	sixty	**sessanta** / *sehs-SAHN-tah*
ten	**dieci** / *dee-EH-chee*	seventy	**settanta** / *seht-TAHN-tah*
eleven	**undici** / *OON-dee-chee*	eighty	**ottanta** / *oht-TAHN-tah*
twelve	**dodici** / *DOH-dee-chee*	ninety	**novanta** / *noh-VAHN-tah*
thirteen	**tredici** / *TREH-dee-chee*	one hundred	**cento** / *CHEHN-to*
fourteen	**quattordici** / *KWAHT-tohr-dee-chee*	two hundred	**duecento** / *DOO-eh-CHEHN-toh*
fifteen	**quindici** / *KWEEN-dee-chee*	one thousand	**mille** / *MEEL-leh*
sixteen	**sedici** / *SEH-dee-chee*		

FRACTIONS & DECIMALS

one eighth	**un ottavo** *oon oht-TAH-voh*
one quarter	**un quarto** *oon KWAHR-toh*
three eighths	**tre ottavi** *treh oht-TAH-vee*
one third	**un terzo** *oon TEHR-tsoh*
one half	**mezzo** *MEHD-dzoh*
two thirds	**due terzi** *DOO-eh TEHR-tsee*
three quarters	**tre quarti** *treh KWAHR-tee*
double	**doppio** *DOHP-pyoh*
triple	**triplo** *TREE-ploh*
one-tenth	**un decimo** *oon DEH-chee-moh*
one-hundredth	**un centesimo** *oon chehn-TEH-zee-moh*
one-thousandth	**un millesimo** *oon meel-LEH-zee-moh*

MATH

addition	**addizione** *ahd-deet-TSYOH-neh*
2 + 1	**due più uno** *DOO-eh PYOO OO-noh*
subtraction	**sottrazione** *soht-traht-TSYOH-neh*
2 − 1	**due meno uno** *DOO-eh MEH-noh OO-noh*

multiplication	**moltiplicazione**
	mohl-tee-plee-kaht-TSYOH-neh
2 x 3	**due per tre**
	DOO-eh pehr treh
division	**divisione**
	dee-vee-ZYOH-neh
6 ÷ 3	**sei diviso tre**
	SEH-ee dee-VEE-zoh treh

ORDINAL NUMBERS

first	**primo -a**
	PREE-moh -mah
second	**secondo -a**
	seh-KOHN-doh -dah
third	**terzo -a**
	TEHR-tsoh -tsah
fourth	**quarto -a**
	KWAHR-toh -tah
fifth	**quinto -a**
	KWEEN-toh -tah
sixth	**sesto -a**
	SEHS-toh -tah
seventh	**settimo -a**
	SEHT-tee-moh -mah
eighth	**ottavo -a**
	oht-TAH-voh -vah
ninth	**nono -a**
	NOH-noh -nah
tenth	**decimo -a**
	DEH-chee-moh -mah
last	**ultimo -a**
	OOL-tee-moh -mah

MEASUREMENTS

Measurements are usually metric, though you may need a few Imperial measurement terms.

centimeter	**centimetro** *chehn-TEE-meh-troh*
meter	**metro** *MEH-troh*
kilometer	**chilometro** *kee-LOH-meh-troh*
millimeter	**millimetro** *meel-LEE-meh-troh*
hectares	**ettari** *EHT-tah-ree*
a distance squared is	**quadrato** *kwah-DRAH-toh*
short	**corto -a** *KOHR-toh -tah*
long	**lungo -a** *LOON-goh -gah*

VOLUME

milliliters	**millilitro** *meel-LEE-lee-troh*
liter	**litro** *LEE-troh*
cup	**tazza** *TAHT-sah*

QUANTITY

some (always singular)	**qualche** *KWAHL-keh*
none	**niente / nessuno** *NYEHN-teh / nehs-SOO-noh*

Dos and Don'ts

Italians measure foodstuffs by the kilogram or smaller 100g unit (**ettogrammo** abbreviated to *etto*: equivalent to just under 4oz). **Pizzerie al taglio** (pizza slice shops) generally run on this system, but hand gestures can suffice. A good server poises the knife, then asks for approval before cutting. **Più** (PYOO) is "more," **meno** (MEH-noh) "less." **Basta** (BAHS-tah) means "enough." To express "half" of something, say **mezzo** (MEHD-zoh).

all	**tutto -a / tutti -e** *TOOT-toh -tah / TOOT-tee -teh*
much / many	**molto -a / molti -e** *MOHL-toh -tah / MOHL-tee -teh*
A little bit (can be used for quantity or for time)	**poco -a** *POH-koh -kah*
dozen	**dozzina** *dot-SEE-nah*

SIZE

small	**piccolo -a** *PEEK-koh-loh-lah*
the smallest (literally "the most small")	**il / la più piccolo -a** *eel / lah PYOO PEEK-koh-loh -lah*
medium	**medio -a** *MEH-dyoh -dyah*
big	**grande** *GRAHN-deh*
fat	**grasso -a** *GRAHS-soh -sah*
really fat	**molto grasso -a** *MOHL-toh GRAHS-soh -sah*

the biggest	**il / la più grande**
	eel / lah PYOO GRAHN-deh
wide	**largo -a**
	LAHR-goh -gah
narrow	**stretto -a**
	STREHT-toh -tah

TIME

Time in Italian is referred to, literally, by the hour. **Che ora è?** translates as "What's the hour?"

For full coverage of number terms, see p7.

HOURS OF THE DAY

What time is it?	**Che ora è?**
	keh OH-rah EH
At what time?	**A che ora?**
	ah keh OH-rah
For how long?	**Per quanto tempo?**
	pehr KWAHN-toh TEHM-poh

A little tip

By adding a diminutive suffix, **-ino -a**, **-etto -a**, or a combination of the two, you can make anything smaller or shorter. These endings replace the original **-o** and **-a**, respectively:

| a really little bit | **pochino -a** (*poh-KEE-noh -nah*) |
| a really teeny tiny bit | **pochettino -a** (*poh-keht-TEE-noh -nah*) |

SURVIVAL ITALIAN

It's one o'clock.	**È l'una.**
	EH LOO-nah
It's two o'clock.	**Sono le due.**
	SOH-noh leh DOO-eh
It's two thirty.	**Sono le due e mezzo.**
	SOH-noh leh DOO-eh eh
	MEHD-dzoh
It's two fifteen.	**Sono le due e un quarto.**
	SOH-noh leh DOO-eh eh oon
	KWAHR-toh
It's a quarter to three.	**Sono le tre meno un quarto. /**
	Manca un quarto alle tre.
	SOH-noh leh treh MEH-noh oon
	KWAHR-toh / MAHN-kah oon
	KWAHR-toh AHL-leh treh
It's noon.	**È mezzogiorno.**
	EH mehd-dzoh -JOHR-noh
It's midnight.	**È mezzanotte.**
	EH mehd-dzah-NOHT-teh
It's early.	**È presto.**
	EH PREHS-toh
It's late.	**È tardi.**
	EH TAHR-dee
in the morning	**al mattino**
	ahl maht-TEE-noh
in the afternoon	**nel pomeriggio**
	nehl poh-meh-REED-joh
at night	**di notte**
	dee NOHT-teh
dawn	**l'alba**
	LAHL-bah

DAYS OF THE WEEK

Sunday	**domenica**
	doh-MEH-nee-kah
Monday	**lunedì**
	loo-neh-DEE
Tuesday	**martedì**
	mahr-teh-DEE

Wednesday	**mercoledì**
	mehr-koh-leh-DEE
Thursday	**giovedì**
	joh-veh-DEE
Friday	**venerdì**
	veh-nehr-DEE
Saturday	**sabato**
	SAH-bah-toh
today	**oggi**
	OHD-jee
tomorrow	**domani**
	doh-MAH-nee
yesterday	**ieri**
	YEH-ree
the day before yesterday	**l'altro ieri**
	lahl-troh-YEH-ree
these last few days	**questi ultimi giorni**
	KWEHS-tee OOL-tee-mee
	JOHR-nee
one week	**una settimana**
	OO-nah seht-tee-MAH-nah
next week	**la prossima settimana**
	lah PROHS-see-mah seht-tee-MAH-nah
last week	**la settimana scorsa**
	lah seht-tee-MAH-nah SKOHR-sah

MONTHS OF THE YEAR

January	**gennaio**
	jehn-NAH-yoh
February	**febbraio**
	fehb-BRAH-yoh
March	**marzo**
	MAHR-tso
April	**aprile**
	ah-PREE-leh
May	**maggio**
	MAHD-joh
June	**giugno**
	JEWN-nyo

July	**luglio**
	LOOL-lyo
August	**agosto**
	ah-GOHS-toh
September	**settembre**
	seht-TEHM-breh
October	**ottobre**
	oht-TOH-breh
November	**novembre**
	noh-VEHM-breh
December	**dicembre**
	dee-CHEHM-breh
one month	**un mese**
	oon MEH-zeh
next month	**il mese prossimo**
	eel MEH-zeh PROHS-see-moh
last month	**il mese scorso**
	eel MEH-zeh SKOHR-soh

SEASONS OF THE YEAR

spring	**la primavera**
	lah pree-mah-VEH-rah
summer	**l'estate**
	lehs-TAH-teh
autumn	**l'autunno**
	low-TOON-noh
winter	**l'inverno**
	leen-VEHR-noh

WEATHER

Weather	**il tempo**
	eel TEHM-poh
What's the weather like?	**Che tempo fa?**
	Kee TEHM-poh fah?
What's the temperature?	**Quanti gradi ci sono?**
	KWAN-tee GRAH-dee chee SOH-noh?
What's the forecast?	**Come sono le previsioni?**
	KOU-meeh SOH-noh leh PREH-veeh-sioh-nee?

ITALIAN GRAMMAR BASICS

Classified as a Romance language, Italian is closely related to Latin, French, Spanish, Portuguese, and Romanian. It developed from the Vulgar Latin of the late Roman Empire.

THE ALPHABET

The Italian alphabet has 21 letters. The letters **h**, **j**, **k**, **w**, **x**, and **y** are used only in words taken from other languages (such as jazz). Certain combinations of letters have special sounds in Italian, just as **ch**, **sh**, **th**, and **ng** have special sounds in English.

Letter	Name	Pronunciation
a	a	**ah** as in *father* - **La Scala** *lah SKAH-lah*
b	bi	**b** as in *bud* - **bacio** *BAH-choh* (kiss)
c	ci	**ca, co, cu, che, chi:** hard **k** sound as in *car* - **cane** *KAH-neh* (dog)
		ce, ci: **ch** as in *cheap* - **ciao** *CHAH-oh*
d	di	**d** as in *day* - **dente** *DEHN-teh* (tooth)
e	e	**eh** as in *bell* - **bene** *BEH-neh* (good, well)
f	effe	**f** as in *fan* - **forte** *FOHR-teh* (strong, loud)
g	gi	**ga, go, gu, ghe, ghi:** hard **g** as in *good* - **gatto** *GAHT-toh* (cat)
		ge, gi: soft **j** as in *jelly* - **gelato** *jeh-LAH-toh* (ice cream)
		gli: **lly** close to *million* - **figlio** *FEEL-lyoh* (son)
		gn: **ny** as in *poignant* - **bagno** *BAHN-nyoh* (bathroom, restroom)
h	acca	silent before vowels, stresses **a** and **o**: **ho** *OH* (I have), **hai** *AH-ee* (you have)
i	i	**ee** as in *eel* - **ieri** *ee-EH-ree* (yesterday)
l	elle	**l** as in *lunch* - **letto** *LET-toh* (bed)
m	emme	**m** as in *Mary* - **mano** *MAH-noh* (hand)
n	enne	**n** as in *nail* - **nero** *NEH-roh* (black)
o	o	**oh** as in *pot* - **oggi** *OHD-jee* (today)
p	pi	**p** as in *pet* - **pasta** *PAHS-tah*
q	cu	**q** as in *quick* - **questo** *KWEHS-toh* (this)
r	erre	trilled, as in the Scottish **r** - **Roma**

Letter	Name	Pronunciation
s	esse	s as in *soon* - **sasso** *SAHS-soh* (rock)
		z as in *rose* between vowels - **casa** *KAH-zah* (house, home)
		sce, sci: sh as in *shop* - **pesce** *PEH-sheh* (fish)
		sche, schi: sk as in *skip* - **schifo** *SKEE-foh* (disgust)
t	ti	t as in *tea* - **Torino** *toh-REE-noh* (Turin)
u	u	oo as in *boom* - **uno** *OO-noh* (one)
v	vu	v as in *very* - **Verona** *veh-ROH-nah*
z	zeta	dz or ts as in *mezzo* and *matzo*
		zucchero *DZOOK-keh-roh* (sugar)
		pizza *PEET-sah*

Foreign letters

j	i lunga	j as *jazz*
k	kappa	k sound
w	doppia vu	w sound

PRONUNCIATION

Italian has few pitfalls like silent letters; a word's sound closely resembles its written form. Such straightforward pronunciation makes this melodious language accessible and appealing to even the most casual student.

Often **c** and **g** are stumbling blocks for beginners. Both have a soft sound before **e** or **i**, a hard sound before **a**, **o**, and **u**. Think **cubo** *KOO-boh* (cube) versus **arrivederci** *ahr-ree-veh-DEHR-chee* (bye) and **gala** *GAH-lah* versus **Luigi** *loo-EE-jee*.

Double consonants should be pronounced twice—or lengthened and intensified. English speakers are already familiar with this from phrases such as gra**b b**ag, bla**ck c**at, goo**d d**ay, hal**f f**ull, goo**d j**ob, ho**t t**ea and ki**ds z**one.

Enjoy the language's drama and richness, but don't slip into an operatic parody. Italians often accuse foreigners of doubling all consonants. Yet no native, when listening to an aria, would ever confuse **m'ama** (she loves me) with **mamma** (mom)!

GENDER, ADJECTIVES, MODIFIERS

The ending of an Italian noun reveals its gender (masculine or feminine) and number (singular and plural). Those ending in **o** are generally masculine and become **i** (plural); those ending in **a** are typically feminine and become **e** (plural). Ones that conclude in **e** can be masculine or feminine, and shift to **i** (plural). Adjectives agree in gender and number, and usually follow the nouns.

	Singular	Plural
Masculine	il piatto bianco (the white plate)	i piatti bianchi (the white plates)
	il cane grande (the large dog)	i cani grandi (the large dogs)
Feminine	la pizza calda (hot pizza)	le pizze calde (hot pizzas)
	la carne tenera (tender meat)	le carni tenere (tender meats)

Nouns often are accompanied by a masculine or feminine definite article (the): **il**, **lo**, **la** (singular); **i**, **gli**, **le** (plural). Indefinite articles (a, an, some)—**un**, **una** (singular) and **dei**, **delle** (plural)—must also correspond to the nouns they modify.

The Definite Article ("The")

	Masculine	Feminine
Singular	il cane (the dog) lo stivale (the boot)	la tavola (the table) la rete (the net)
Plural	i cani (the dogs) gli stivali (the boots)	le tavole (the tables) le reti (the nets)

The Indefinite Article ("A" or "An")

	Masculine	Feminine
Singular	un cane (a dog)	una tavola (a table)
Plural	dei cani (some dogs)	delle tavole (some tables)

PERSONAL PRONOUNS

English	Italian	Pronunciation
I	io	EE-oh
You (singular, familiar)	tu	TOO
He / She / You (singular, formal)	lui / lei / Lei	LOO-ee / lay / lay
We	noi	NOH-ee
You (plural, familiar)	voi	VOH-ee
They / You (plural, formal)	loro / Loro	LOH-roh

PRONOUNS

English	Italian	Pronunciation
This	questo -a	KWEHS-toh -tah
That	quello -a	KWEHL-loh -lah
These	questi -e	KWEHS-tee -teh
Those	quelli -e	KWEHL-lee -leh

Hey, You!

Italian has two words for "you"—**tu**, spoken among friends and familiars, and to address children; and **Lei / Loro**, used among strangers or as a sign of respect toward elders and authority figures. When speaking with a stranger, expect to use **Lei / Loro** unless you are invited to do otherwise. These days you will almost always hear the familiar form (**voi**) rather than the formal plural (**loro**) used in speech.

REGULAR VERB CONJUGATIONS

Italian verb infinitives end in **ARE** (e.g. **parlare**, to speak), **ERE** (e.g. **vendere**, to sell), and **IRE** (e.g. **partire**, to leave). Drop the last three letters to determine the word's stem. Then add endings that reveal who did the action—and when. Following are the present-tense conjugations for regular verbs.

Present Tense

ARE Verbs	PARLARE "To Talk, To Speak"	
I talk.	Io parlo.	PAHR-loh
You (singular, familiar) talk.	Tu parli.	PAHR-lee
He / She talks. You (singular, formal) talk.	Lui / Lei / Lei parla.	PAHR-lah
We talk.	Noi parliamo.	pahr-LYAH-moh
You (plural, familiar) talk.	Voi parlate.	pahr-LAH-teh
They / You (plural, formal) talk.	Loro / Loro parlano.	PAHR-lah-noh

ERE Verbs	VENDERE "To Sell"	
I sell.	Io vendo.	VEHN-doh
You (singular, familiar) sell.	Tu vendi.	VEHN-dee
He / She sells. You (singular, formal) sell.	Lui / Lei / Lei vende.	VEHN-deh
We sell.	Noi vendiamo.	vehn-DYAH-moh
You (plural, familiar) sell.	Voi vendete.	vehn-DEH-teh
They / You (plural, formal) sell.	Loro / Loro vendono.	VEHN-doh-noh

IRE Verbs	PARTIRE "To Leave"	
I leave.	Io parto.	PAHR-toh
You (singular, familiar) **leave.**	Tu parti.	PAHR-tee
He / She leaves. You (singular, formal) **leave.**	Lui / Lei / Lei parte.	PAHR-teh
We leave.	Noi partiamo.	pahr-TYAH-moh
You (plural, familiar) **leave.**	Voi partite.	pahr-TEE-teh
They / You (plural, formal) **leave.**	Loro / Loro partono.	PAHR-toh-noh

Past Tense

Italian has five past tenses. The present perfect most often expresses the simple past (equivalent to the English "I have eaten" or "I ate"). The imperfect tense conveys an unfinished or continuing action ("I was eating"). So for verbs like **essere** (to be) we've supplied that form instead. Below are examples for regular verbs.

ARE Verbs	PARLARE "To Talk, To Speak"	
I talked.	Io ho parlato.	oh pahr-LAH-toh
You (singular, familiar) **talked.**	Tu hai parlato.	eye pahr-LAH-toh
He / She / You (singular, formal) **talked.**	Lui / Lei / Lei ha parlato.	AH pahr-LAH-toh
We talked.	Noi abbiamo parlato.	ahb-BYAH-moh pahr-LAH-toh
You (plural, familiar) **talked.**	Voi avete parlato.	ah-VEH-teh pahr-LAH-toh
They / You (plural, formal) **talked.**	Loro / Loro hanno parlato.	AHN-noh pahr-LAH-toh

ERE Verbs

	VENDERE "To Sell"	
I sold.	Io ho venduto.	oh vehn-DOO-toh
You (singular, familiar) sold.	Tu hai venduto.	eye vehn-DOO-toh
He / She / You (singular, formal) sold.	Lui / Lei / Lei ha venduto.	AH vehn-DOO-toh
We sold.	Noi abbiamo venduto.	ahb-BYAH-moh vehn-DOO-toh
You (plural, familiar) sold.	Voi avete venduto.	ah-VEH-teh vehn-DOO-toh
They / You (plural, formal) sold.	Loro / Loro hanno venduto.	AHN-noh vehn-DOO-toh

IRE Verbs

	PARTIRE "To Leave"	
I left.	Io sono partito.	SOH-noh pahr-TEE-toh
You (singular, familiar) left.	Tu sei partito.	SEH-ee pahr-TEE-toh
He / She / You (singular, formal) left.	Lui / Lei / Lei è partito.	EH pahr-TEE-toh
We left.	Noi siamo partiti.	SYAH-moh pahr-TEE-tee
You (plural, familiar) left.	Voi siete partiti.	SYEH-teh pahr-TEE-tee
They / You (plural, formal) left.	Loro / Loro sono partiti.	SOH-noh pahr-TEE-tee

The Future Tense

ARE Verbs | **PARLARE "To Talk, To Speak"**

I will talk.	Io parlerò.	pahr-leh-ROH
You (singular, familiar) will talk.	Tu parlerai.	pahr-leh-REYE
He / she / you (singular, formal) will talk.	Lui / Lei / Lei parlerà.	pahr-leh-RAH
We will talk.	Noi parleremo.	pahr-leh-REH-moh
You (plural, familiar) will talk.	Voi parlerete.	pahr-leh-REH-teh
They / You (plural, formal) will talk.	Loro / Loro parleranno.	pahr-leh-RAHN-noh

ERE Verbs | **VENDERE "To Sell"**

I will sell.	Io venderò.	vehn-deh-ROH
You (singular, familiar) will sell.	Tu venderai.	vehn-deh-REYE
He / She / You (singular, formal) will sell.	Lui / Lei / Lei venderà.	vehn-deh-RAH
We will sell.	Noi venderemo.	vehn-deh-REH-moh
You (plural, familiar) will sell.	Voi venderete.	vehn-deh-REH-teh
They / You (plural, formal) will sell.	Loro / Loro venderanno.	vehn-deh-RAHN-noh

IRE Verbs	PARTIRE "To Leave"	
I will leave.	Io partirò.	pahr-tee-ROH
You (singular, familiar) will leave.	Tu partirai.	pahr-tee-REYE
He / She / You (singular, formal) will leave.	Lui / Lei / Lei partirà.	pahr-tee-RAH
We will leave.	Noi partiremo.	pahr-tee-REH-moh
You (plural, familiar) will leave.	Voi partirete.	pahr-tee-REH-teh
They / You (plural, formal) will leave.	Loro / Loro partiranno.	pahr-tee-RAHN-noh

TO BE OR NOT TO BE

Italian has two verbs that mean "to be" (am, are, is, was, were). One is for physical location or temporary conditions (**stare**), and the other is for fixed qualities or conditions (**essere**). **Stare** is used in courtesy expressions with **bene** or **male** (well, unwell), e.g. **Come sta? Sto bene / male** (How are you? I'm fine / not well) and to express a progressive -ing action, e.g. **Sto mangiando** (I am eating). **Essere** is used most of the time to express fixed qualities or conditions (as in English) and health states other than **bene** or **male**, e.g. **Sono stanco** (I'm tired).

I am here. (temporary, stare)	Io sto qua.
What are you doing? (temporary, stare)	Cosa sta facendo?
The train is slow. (quality, essere)	Il treno è lento.

Stare "To Be, To Stay" (conditional)
Present Tense

I am.	Io sto.	stoh
You (singular, familiar) **are.**	Tu stai.	steye
He / She is. You (singular, formal) **are.**	Lui / Lei / Lei sta.	stah
We are.	Noi stiamo.	STYAH-moh
You (plural, familiar) **are.**	Voi state.	STAH-teh
They / You (plural, formal) **are.**	Loro / Loro stanno.	STAHN-noh

Past (imperfect) Tense

I was.	Io stavo.	STAH-voh
You (singular, familiar **were.**	Tu stavi.	STAH-vee
He / She was. You (singular, formal) **were.**	Lui / Lei / Lei stava.	STAH-vah
We were.	Noi stavamo.	stah-VAH-moh
You (plural, familiar) **were.**	Voi stavate.	stah-VAH-teh
They / You (plural, formal) **were.**	Loro / Loro stavano.	STAH-vah-noh

Essere "To Be" (permanent)

Present Tense

I am.	Io sono.	SOH-noh
You (singular, familiar) **are.**	Tu sei.	SEH-ee
He /She is. You (singular, formal) **are.**	Lui / Lei / Lei è.	EH
We are.	Noi siamo.	SYAH-moh
You (plural, familiar) **are.**	Voi siete.	SYEH-teh
They / You (plural, formal) **are.**	Loro / Loro sono.	SOH-noh

Past (imperfect) Tense

I was.	Io ero.	EH-roh
You (singular, familiar) **were.**	Tu eri.	EH-ree
He / She was. You (singular, formal) **were.**	Lui / Lei / Lei era.	EH-rah
We were.	Noi eravamo.	eh-rah-VAH-moh
You (plural, familiar) **were.**	Voi eravate.	eh-rah-VAH-teh
They / You (plural, formal) **were.**	Loro / Loro erano.	EH-rah-noh

IRREGULAR VERBS

Italian has numerous irregular verbs that stray from the standard -**ARE**, -**ERE**, and -**IRE** conjugations. Rather than bog you down with too much grammar, we're providing the present tense conjugations for the most common irregular verbs.

AVERE "To Have"

I have.	Io ho.	OH
You (singular, familiar) have.	Tu hai.	EYE
He / She has. You (singular, formal) have.	Lui / Lei / Lei ha.	AH
We have.	Noi abbiamo.	ahb-BYAH-moh
You (plural, familiar) have.	Voi avete.	ah-VEH-teh
They / You (plural, formal) have.	Loro / Loro hanno.	AHN-noh

ANDARE "To Go"

I go.	Io vado.	VAH-doh
You (singular, familiar) go.	Tu vai.	VAH-ee
He / She goes. You (singular, formal) go.	Lui / Lei / Lei va.	vah
We go.	Noi andiamo.	ahn-DYAH-moh
You (plural, familiar) go.	Voi andate.	ahn-DAH-teh
They / You (plural, formal) go.	Loro / Loro vanno.	VAHN-noh

DARE "To Give"

I give.	Io do.	DOH
You (singular, familiar) **give.**	Tu dai.	DAH-ee
He / She gives. You (singular, formal) **give.**	Lui / Lei / Lei dà.	DAH
We give.	Noi diamo.	DYAH-moh
You (plural, familiar) **give.**	Voi date.	DAH-teh
They / You (plural, formal) **give.**	Loro / Loro danno.	DAHN-noh

FARE "To Do, To Make"

I do / make.	Io faccio.	FAHT-choh
You (singular, familiar) **do / make.**	Tu fai.	FAH-ee
He / She does / makes. You (singular, formal) **do / make.**	Lui / Lei / Lei fa.	FAH
We do / make.	Noi facciamo.	faht-CHAH-moh
You (plural, familiar) **do / make.**	Voi fate.	FAH-teh
They / You (plural, formal) **do / make.**	Loro / Loro fanno.	FAHN-noh

Fare

The verb *fare* means to make or do. It's also used to describe the weather. For example:

Fa caldo. It's hot.
 (Literally: It makes hot.)
Fa freddo. It's cold.
 (Literally: It makes cold.)

Be careful not to say **Sono freddo**, as this translates "I'm a cold person" or "I'm sexually frigid". Instead, say **Ho freddo / caldo.** (Literally: I have cold / hot.) Likewise, **Sono caldo** can mean "hot to trot."

	BERE "To Drink"	
I drink.	Io bevo.	BEH-voh
You (singular, familiar) **drink.**	Tu bevi.	BEH-vee
He / She drinks. You (singular, formal) **drink.**	Lui / Lei / Lei beve.	BEH-veh
We drink.	Noi beviamo.	beh-VYAH-moh
You (plural, familiar) **drink.**	Voi bevete.	beh-VEH-teh
They / You (plural, formal) **drink.**	Loro / Loro bevono.	BEH-voh-noh

DOVERE "Must, To Have To"

I must / have to.	**Io devo.**	DEH-voh
You (singular, familiar) **must / have to.**	**Tu devi.**	DEH-vee
He / She must / has to. You (singular, formal) **must / have to.**	**Lui / Lei / Lei deve.**	DEH-veh
We must / have to.	**Noi dobbiamo.**	dohb-BYAH-moh
You (plural, familiar) **must / have to.**	**Voi dovete.**	doh-VEH-teh
They / You (plural, formal) **must / have to.**	**Loro / Loro devono.**	DEH-voh-noh

POTERE "Can, To Be Able"

I can / am able.	**Io posso.**	POHS-soh
You (singular, familiar) **can / are able.**	**Tu puoi.**	poo-OH-ee
He / She can/ is able. You (singular, formal) **can / are able.**	**Lui /Lei / Lei può.**	poo-OH
We can / are able.	**Noi possiamo.**	pohs-SYAH-moh
You (plural, familiar) **can / are able.**	**Voi potete.**	poh-TEH-teh
They / You (plural, formal) **can / are able.**	**Loro / Loro possono.**	POHS-soh-noh

SAPERE "To Know"

I know.	Io so.	soh
You (singular, familiar) know.	Tu sai.	SAH-ee
He / She knows. You (singular, formal) know.	Lui / Lei / Lei sa.	SAH
We know.	Noi sappiamo.	sahp-PYAH-moh
You (plural, familiar) know.	Voi sapete.	sah-PEH-teh
They / You (plural, formal) know.	Loro / Loro sanno.	SAHN-noh

VOLERE "To Want"

I want.	Io voglio.	VOHL-lyoh
You (singular, familiar) want.	Tu vuoi.	VWOH-ee
He / She wants. You (singular, formal) want.	Lui / Lei / Lei vuole.	VWOH-leh
We want.	Noi vogliamo.	vohl-LYAH-moh
You (plural, familiar) want.	Voi volete.	voh-LEH-teh
They / You (plural, formal) want.	Loro / Loro vogliono.	VOHL-lyoh-noh

USCIRE "To Get Out"

I get out.	Io esco.	EHS-koh
You (singular, familiar) **get out.**	Tu esci.	EH-shee
He / She gets out. You (singular, formal) **get out.**	Lui / Lei / Lei esce.	EH-sheh
We get out.	Noi usciamo.	oo-SHAH-moh
You (plural, familiar) **get out.**	Voi uscite.	oo-SHEE-teh
They / You (plural, formal) **get out.**	Loro / Loro escono.	EHS-koh-noh

VENIRE "To Come"

I come.	Io vengo.	VEHN-goh
You (singular, familiar) **come.**	Tu vieni.	VYEH-nee
He / She comes. You (singular, formal) **come.**	Lui / Lei / Lei viene.	VYEH-neh
We come.	Noi veniamo.	veh-NYAH-moh
You (plural, familiar) **come.**	Voi venite.	veh-NEE-teh
They / You (plural, formal) **come.**	Loro / Loro vengono.	VEHN-goh-noh

Piacere

The Italian word for "to like" is *piacere*, which literally means to please. So, rather than "I like chocolate," Italians say, "Chocolate is pleasing to me."

Mi piace il cioccolato. I like chocolate.
  (Literally: Chocolate is pleasing to me.)
Mi piacciono i dolci. I like sweets / desserts.
  (Literally: Sweets are pleasing to me.)

Piacere "To Like"

Present Tense	Singular	Plural
I like.	Mi piace. PYAH-cheh	Mi piacciono. PYAHT-choh-noh
You (informal, singular) like.	Ti piace.	Ti piacciono.
He / She likes. You (formal, singular) like.	Gli / Le / Le piace.	Gli / Le / Le piacciono.
We like.	Ci piace.	Ci piacciono.
You (informal, plural) like.	Vi piace.	Vi piacciono.
They / You (formal, plural) like.	Gli / Loro piace.	Gli / Loro piacciono.

Past Tense	Singular	Plural
I liked.	**Mi piaceva.** pyah-CHEH-vah	**Mi piacevano.** pyah-CHEH-vah-noh
You (informal, singular) **liked.**	**Ti piaceva.**	**Ti piacevano.**
He / She/ You (formal, singular) **liked.**	**Gli / Le / Le piaceva.**	**Gli / le / Le piacevano.**
We liked.	**Ci piaceva.**	**Ci piacevano.**
You (informal, plural) **liked.**	**Vi piaceva.**	**Vi piacevano.**
They / You (formal, plural) **liked.**	**Gli / Loro piaceva.**	**Gli / Loro piacevano.**

REFLEXIVE VERBS

Italian has many more reflexive verbs than English. A verb is reflexive when both its subject and object refer to the same person or thing. For example: "Maria looks at herself in the mirror," **Maria si guarda allo specchio**. The following common verbs are used reflexively: **vestirsi** (to get dressed, literally to dress oneself), **bagnarsi** (to get oneself wet), and **svegliarsi** (to wake up, literally to wake oneself up).

VESTIRSI "To Get Dressed"

I get dressed.	Io mi vesto.	mee VEHS-toh
You (singular, familiar) get dressed.	Tu ti vesti.	tee VEHS-tee
He / She gets dressed. You (singular, formal) get dressed.	Lui / Lei / Lei si veste.	see VEHS-teh
We get dressed.	Noi ci vestiamo.	chee vehs-TYAH-moh
You (plural, familiar) get dressed.	Voi vi vestite.	vee vehs-TEE-teh
They / You (plural, formal) get dressed.	Loro / Loro si vestono.	see VEHS-toh-noh

CHAPTER TWO

GETTING THERE & GETTING AROUND

This section deals with every form of transportation. Whether you've just reached your destination by plane or you're renting a car to tour the countryside, you'll find the necessary phrases in the next 30 pages.

AT THE AIRPORT

I am looking for _____	**Cerco** _____ *CHEHR-koh*
a bus/train to city center.	**Un autobus / treno per il centro.** *oon ah-oo-TOH-bus / TREH-noh pehr eel CEHN-troh*
a porter.	**un facchino.** *oon fahk-KEE-noh*
the check-in counter.	**il check-in.** *eel check-in*
the ticket counter.	**la biglietteria.** *lah beel-lyeht-teh-REE-ah*
security.	**il controllo sicurezza.** *eel kon-TROLL-loh see-kuh-REHT-tzah*
immigration.	**l'ufficio immigrazione.** *loof-FEE-choh eem-mee-grah-TZYOH-neh*
customs.	**la dogana.** *lah doh-GAH-nah*
arrivals.	**gli arrivi.** *lyee ahr-REE-vee*
departures.	**le partenze.** *leh pahr-TEHN-tseh*
gate number _____.	**l'uscita numero _____.** *loo-SHEE-tah NOO-meh-roh*

For full coverage of numbers, see p7.

the waiting area.	**la sala d'attesa.** *lah SAH-lah daht-TEH-zah*
the men's restroom.	**il bagno degli uomini.** *eel BAHN-nyo DEHL-lyee WOH-mee-nee*
the women's restroom.	**il bagno delle donne.** *eel BAHN-nyoh dehl-leh DOHN-neh*
the police station.	**la stazione di polizia.** *lah stah-TSYOH-neh dee poh-lee-TSEE-ah*
a security guard.	**una guardia di sicurezza.** *OOH-nah GWAHR-dyah dee see-koo-RET-sah*
the smoking area.	**l'area fumatori.** *LAH-reh-ah foo-mah-TOH-ree*
the information booth.	**il punto informazioni.** *eel POON-toh een-FOHR-mah-TSYOH-nee*
a public telephone.	**un telefono pubblico.** *oon teh-LEH-foh-noh POOB-blee-koh*
an ATM / cashpoint.	**un bancomat.** *oon BAHN-koh-maht*
baggage claim.	**il ritiro bagagli.** *eel ree-TEE-roh bah-GAHL-lyee*
a luggage cart.	**un carrello.** *oon kahr-REHL-loh*
a currency exchange.	**un cambiavalute.** *oon KAHM-byah-vah-LOO-teh*
a café.	**un caffè.** *oon kahf-FEH*
a restaurant.	**un ristorante.** *oon ree-stoh-RAHN-teh*
a bar.	**un bar.** *oon bar*

GETTING THERE

a bookstore or newsstand.	**una libreria o un'edicola.** *OO-nah lee-breh-REE-ah oh oon eh-DEE-koh-lah*
a duty-free shop.	**un duty-free.** *oon duty-free*
Is there Wi-Fi here?	**C'è il Wi-Fi qui?** *ch-EH eel Wi-Fi kwee*
I'd like to page someone.	**Vorrei far chiamare qualcuno.** *vohr-RAY fahr kyah-MAH-reh kwahl-KOO-noh*
Do you accept credit cards?	**Accettate la carta di credito?** *ah-cheht-TAH-teh lah KAHR-tah dee KREH-dee-toh*

CHECKING IN

I would like a one-way ticket to _____.	**Vorrei un biglietto di andata per _____.** *vohr-RAY oon beel-LYEHT-toh dee ahn-DAH-tah pehr*
I would like a round trip ticket to _____.	**Vorrei un biglietto di andata e ritorno per _____.** *vohr-RAY oon beel-LYEHT-toh dee ahn-DAH-tah eh ree-TOHR-noh pehr*
How much are the tickets?	**Quanto costano i biglietti?** *KWAHN-toh KOHS-tah-noh ee beel-LYEHT-tee*
Do you have anything less expensive?	**C'è qualcosa di meno caro?** *ch-EH kwahl-KOH-zah dee MEH-noh KAH-roh*
What time does flight _____ leave?	**A che ora parte il volo _____?** *ah keh OH-rah PAHR-teh eel VOH-loh*
What time does flight _____ arrive?	**A che ora arriva il volo _____?** *ah keh OH-rah ahr-REE-vah eel VOH-loh*

For full coverage of numbers, see p7.
For full coverage of time, see p12.

Common Airport Signs

Arrivi	Arrivals
Partenze	Departures
Terminal	Terminal
Uscita	Gate
Emissione biglietti	Ticketing
Dogana	Customs
Ritiro bagagli	Baggage Claim
Spingere	Push
Tirare	Pull
Vietato fumare	No Smoking
Entrata	Entrance
Uscita	Exit
Uomini	Men's
Donne	Women's
Bus navetta	Shuttle Buses
Taxi	Taxis

GETTING THERE

How long is the flight?

Quanto dura il volo?
KWAHN-toh DOO-rah eel VOH-loh

Do I have a connecting flight?

C'è una coincidenza?
*ch-EH OO-nah
koh-een-chee-DEHN-tsa*

Do I need to change planes?

Devo cambiare aereo?
*DEH-voh kahm-BYAH-reh
ah-EH-reh-oh*

My flight leaves at __:__.

Il mio aereo parte alle __:__.
*eel MEE-oh ah-EH-reh-oh
PAHR-teh ahl-leh*

What time will the flight arrive?

A che ora arriva l'aereo?
*ah keh OH-rah ahr-REE-vah lah-
EH-reh-oh*

Is the flight on time?	**Il volo è in orario?** *eel VOH-loh EH een oh-RAH-ryoh*
Is the flight delayed?	**Il volo è in ritardo?** *eel VOH-loh EH een ree-TAHR-doh*
From which terminal is flight ____ leaving?	**Da che terminal parte il volo ____?** *dah keh TEHR-mee-nahl PAHR-teh eel VOH-loh*
From which gate is flight ____ leaving?	**Da che uscita parte il volo ____?** *dah keh oo-SHEE-tah PAHR-teh eel VOH-loh*

For full coverage of numbers, see p7.

How much time do I need for check-in?	**Quanto tempo ci vuole per fare il check-in?** *KWAHN-toh TEHM-poh chee VWOH-leh pehr FAH-reh eel check-in*
Is there an express check-in line?	**C'è una fila per il check-in espresso?** *che-EH-OO-nah FEE-lah pehr eel check-in ess-press-soh*
Is online check-in available?	**È possibile fare il check-in on-line?** *EH pohs-SEE-bee-leh FAH-reh eel check-in on-line*

Questions you may be asked

Il passaporto, per favore.
eel pahs-sah-POHR-toh pehr fah-VOH-reh

Your passport, please.

Qual'è lo scopo del suo viaggio?
kwah-LEH loh SKOH-poh dehl suoh VYAHD-joh

What is the purpose of your visit?

Per quanto tempo si ferma?
Pehr KWAHN-toh TEHM-poh see FEHR-mah

How long will you be staying?

Dove alloggerà?
DOH-veh ahl-loh-JEH-rah

Where are you staying?

Ha qualcosa da dichiarare?
ah kwahl-KOH-zah dah dee-kyah-RAH-reh

Do you have anything to declare?

Apra questa borsa, per favore.
ah-PRAH KWEHS-tah BOHR-sah pehr fah-VOH-reh

Open this bag, please.

GETTING THERE

Seat Preferences

I would like _____ ticket(s) in _____

Vorrei _____ biglietto -i in _____
vohr-RAY _____ beel-LYEHT-toh -ee een _____

first / business class.

prima classe / Business class.
PREE-mah KLAHS-seh

economy class.

classe turistica.
KLAHS-seh too-REES-tee-kah

I would like _____

Vorrei _____
vohr-RAY _____

Please don't give me _____

Per favore non mi dia _____
pehr fah-VOH-reh nohn mee DEE-ah _____

a window seat.

un posto vicino al finestrino.
oon POHS-toh vee-CHEE-noh ahl fee-nehs-TREE-noh

an aisle seat.

un posto sul corridoio.
oon POHS-toh sool kohr-ree-DOH-yoh

an emergency exit row seat.

un posto vicino all'uscita di sicurezza.
oon POHS-toh vee-CHEE-noh ahl-loo-SHEE-tah dee see-koo-RET-sah

a bulkhead seat.

un posto in prima fila.
oon POHS-toh een PREE-mah FEE-lah

a seat by the restroom.	**un posto vicino alla toilette.** *oon POHS-toh vee-CHEE-noh AHL-leh twa-LEHT*
a seat near the front.	**un posto nella parte anteriore.** *oon POHS-toh NEHL-lah PAHR-teh ahn-teh-RYOH-reh*
a seat near the middle.	**un posto nella zona centrale.** *oon POHS-toh NEHL-lah DZOH-nah chehn-TRAH-leh*
a seat near the back.	**un posto nella zona posteriore.** *oon POHS-toh nehl-lah ZOH-nah poh-steh-RYOH-reh*
Is there a meal on the flight?	**Viene servito un pasto durante il volo?** *VYEH-neh sehr-VEE-toh oon PAHS-toh doo-RAHN-teh eel VOH-loh*
I'd like to order _____	**Vorrei ordinare _____** *vohr-RAY ohr-dee-NAH-reh*
a vegetarian meal.	**un pasto vegetariano.** *oon PAHS-toh veh-jeh-tah-RYAH-noh*
a gluten-free meal.	**un pasto senza glutine.** *ooh PAHS-toh SEHN-tsah gloo-TEE-neh*
a kosher meal.	**un pasto kasher.** *oon PAHS-toh KAH-shehr*
a diabetic meal.	**un pasto per diabetici.** *oon PAHS-toh pehr dyah-BEH-tee-chee*
I am traveling to _____.	**Sto andando a _____.** *stoh ahn-DAHN-doh ah*
I am coming from _____.	**Vengo da _____.** *VEHN-goh dah*
I arrived from _____.	**Arrivo da _____.** *ahr-REE-voh dah*

For full coverage of country terms, see English / Italian dictionary.

I'd like to change / cancel / confirm my reservation.	**Vorrei cambiare / annullare / confermare la mia prenotazione.** *vohr-RAY kahm-BYAH-reh / ahn-nool-LAH-reh / kohn-fehr-MAH-reh lah MEE-ah preh-noh-tah-TSYOH-neh*
I have ___ bags to check.	**Ho ___ bagagli da registrare.** *OH ___ bah-GAHL-lyee dah reh-jees-TRAH-reh*

For full coverage of numbers, see p7.

Passengers with Special Needs

Is that handicap accessible?	**C'è accesso ai disabili?** *ch-EH atch-CHESS-oh eye dee-ZAH-bee-lee*
May I have a wheelchair / a walker please?	**Posso avere una sedia a rotelle / un deambulatore, per favore?** *POHS-soh ah-VEH-reh OO-nah SEH-dyah ah roh-TEHL-leh / oon deh-ahm-boo-lah-TOH-reh pehr fah-VOH-reh*
I need some assistance boarding.	**Ho bisogno di assistenza all'imbarco.** *OH bee-ZOHN-nyoh dee ahs-sees-TEHN-tsa ahl-leem-BAHR-koh*
I need to bring my service dog.	**Devo portare il mio cane guida.** *DEH-voh pohr-TAH-reh eel MEE-oh KAH-neh GWEE-dah*
Do you have services for the hearing impaired?	**Ci sono servizi per ipoudenti?** *chee SOH-noh sehr-VEE-tsee pehr EE-poh-oo-DEHN-tee*
Do you have services for the visually impaired?	**Ci sono servizi per ipovedenti?** *chee SOH-noh sehr-VEE-tsee pehr EE-poh-oo-veh-DEHN-tee*

Trouble at Check-In

How long is the delay?	**Di quanto è il ritardo?**
	dee KWAHN-toh EH eel ree-TAHR-doh
My flight was late.	**Il mio volo era in ritardo.**
	eel MEE-oh VOH-loh EH-rah een
	ree-TAHR-doh
I missed my flight.	**Ho perso il volo.**
	OH PEHR-soh eel VOH-loh
When is the next flight?	**Quand'è il prossimo volo?**
	kwahn-DEH eel PROHS-see-moh
	VOH-loh
May I have a meal voucher?	**Posso avere un buono pasto?**
	POHS-soh ah-VEH-reh oon BWOH-
	noh PAHS-toh
May I have a room voucher?	**Posso avere un buono stanza?**
	POHS-soh ah-VEH-reh oon BWOH-
	noh STAHN-tsah

AT CUSTOMS / SECURITY CHECKPOINTS

I'm traveling with a group.	**Viaggio con un gruppo.**
	VYAHD-joh kohn oon GROOP-poh
I'm on my own.	**Viaggio da solo -a.**
	VYAHD-joh dah SOH-loh -ah
I'm traveling on business.	**Sono in viaggio per lavoro.**
	SOH-noh een VYAHD-joh
	pehr lah-VOH-roh
I'm on vacation.	**Sono in vacanza.**
	SOH-noh een vah-KAHN-tsah
I have nothing to declare.	**Non ho nulla da dichiarare.**
	nohn OH NOOL-lah dah
	dee-kyah-RAH-reh
I would like to declare ____.	**Vorrei dichiarare ____.**
	vohr-RAY dee-kyah-RAH-reh
I have some liquor.	**Ho del liquore.**
	OH dehl lee-KWOH-reh
I have some cigars.	**Ho dei sigari.**
	OH day SEE-gah-ree

They are gifts.	**Sono regali.**
	SOH-noh reh-GAH-lee
They are for personal use.	**Sono ad uso personale.**
	SOH-noh ahd OO-zoh pehr-soh-NAH-leh
That is my medicine.	**È la mia medicina.**
	EH lah MEE-ah meh-dee-CHEE-nah
I have my prescription.	**Ho la mia ricetta.**
	OH lah MEE-ah ree-CHET-tah
I'd like a female / male officer to conduct the search.	**Vorrei un agente donna / uomo per la perquisizione.**
	vohr-RAY oon ah-JEN-teh DOHN-nah / WOH-moh pehr lah pehr-kwee-zee-TSYOH-neh

Listen Up: Security Lingo

Per favore, ____	Please ____
pehr fah-VOH-reh	
si tolga le scarpe.	remove your shoes.
see TOHL-gah leh SKAHR-peh	
si tolga la giacca / la maglia.	remove your jacket / sweater.
see TOHL-gah lah JAHK-kah / lah MAH-lyah	
si tolga i gioielli.	remove your jewelry.
see TOHL-gah ee joh-YEHL-lee	
metta le borse sul trasportatore.	place your bags on the conveyor belt.
MEHT-tah leh BOHR-seh SOOL trah-spohr-tah-TOH-reh	
Si sposti di qua.	Step to the side.
see SPOH-stee dee KWAH	
Dobbiamo fare un'ispezione manuale.	We have to do a hand search.
dohb-BYAH-moh FAH-reh oon ee-speh-TSYOH-neh mah-NWAH-leh	

Trouble at Security

Help me. I've lost _____	**Mi aiuti. Ho perso _____**
	mee ah-YOO-tee OH PEHR-soh
my passport.	**il passaporto.**
	eel pahs-sah-POHR-toh
my boarding pass.	**la carta d'imbarco.**
	lah KAHR-tah deem-BAHR-koh
my identification.	**il documento d'identità.**
	eel doh-koo-MEHN-toh
	dee-dehn-tee-TAH
my wallet.	**il portafoglio.**
	eel pohr-tah-FOHL-lyoh
my purse.	**la borsa.**
	lah BOHR-sah
Someone stole my purse / wallet!	**Mi hanno rubato la borsa / il portafoglio!**
	mee AHN-noh roo-BAH-toh lah BOHR-sah / eel pohr-tah-FOHL-lyoh

IN-FLIGHT

It's unlikely you'll need much Italian on the plane, but these phrases will help if a bilingual flight attendant is unavailable or if you need to talk to an Italian-speaking neighbor.

I think that's my seat.	**Credo che quello sia il mio posto.**
	KREH-doh keh KWEHL-loh SEE-ah eel MEE-oh POHS-toh
May I have _____	**Posso avere _____**
	POHS-soh ah-VEH-reh
water?	**dell'acqua?**
	dehl-LAHK-wah
sparkling water?	**dell'acqua gassata?**
	dehl-LAHK-wah gas-SAH-tah
orange juice?	**del succo d'arancia?**
	dehl SOOK-koh dah-RAHN-chah

soda?	**una bibita?**
	OO-nah BEE-bee-tah
diet soda?	**una bibita light?**
	OO-nah BEE-bee-tah light
a beer?	**una birra?**
	OO-nah BEER-rah
wine?	**del vino?**
	dehl VEE-noh

For a complete list of drinks, see p88.

a pillow?	**un cuscino?**
	oon koo-SHEE-noh
a blanket?	**una coperta?**
	OO-nah koh-PEHR-tah
headphones?	**le cuffie?**
	leh KOOF-fyeh
a magazine or newspaper?	**una rivista o un giornale?**
	OO-nah ree-VEES-tah oh oon johr-NAH-leh

GETTING THERE

When will the meal be served?
Quando sarà servito il pasto?
KWAHN-doh sah-RAH sehr-VEE-toh eel PAHS-toh

How long until we land?
Quanto manca all'atterraggio?
KWAHN-toh MAHN-kah ahl-laht-tehr-RAHD-joh

May I move to another seat?
Posso cambiare posto?
POHS-soh kahm-BYAH-reh POHS-toh

How do I turn the light on / off?
Come si accende / spegne la luce?
KOH-meh see atch-CHEN-deh / SPEHN-nyeh lah LOO-cheh

Trouble In-Flight
These headphones are broken.
Queste cuffie sono guaste.
KWEHS-teh KOOF-fyeh SOH-noh GWAHS-teh

I spilled.	**Mi si è rovesciato.** *mee see EH roh-veh-SHAH-toh*
My child spilled.	**Il bambino ha rovesciato.** *eel bahm-BEE-noh ah roh-veh-SHAH-toh*
My child is sick.	**Il mio bambino / la mia bambina non sta bene.** *eel MEE-oh bahm-BEE-noh / lah MEE-ah bahm-BEE-nah non stah bay-nay*
I need an airsickness bag.	**Mi serve un sacchetto per il mal d'aria.** *mee SEHR-veh oon sahk-KEHT-toh perh eel mahl DAHR-yah*
I smell something strange.	**Sento uno strano odore.** *SEHN-toh OO-noh STRAH-noh oh-DOH-reh*
That passenger is behaving suspiciously.	**Quel passeggero si comporta in modo sospetto.** *kwehl pahs-sed-JEH-roh see kohm-POHR-tah een MOH-doh sohs-PEHT-to*

BAGGAGE CLAIM

Where is baggage claim for flight _____?	**Dov'è il ritiro bagagli per il volo _____?** *doh-VEH eel ree-TEE-roh bah-GAHL-lyee perh eel VOH-loh*
Would you please help with my bags?	**Può aiutarmi con i bagagli?** *PWOH ah-yoo-TAHR-mee kohn ee bah-GAHL-lyee*
I am missing _____ bags.	**Mi mancano _____ borse.** *mee MAHN-kah-noh _____ BOHR-seh*

For full coverage of numbers, see p7.

My bag is _____	**La mia borsa è _____**
	lah MEE-ah BOHR-sah EH
lost.	**smarrita.**
	zmahr-REE-tah
damaged.	**danneggiata.**
	dahn-ned-JAH-tah
stolen.	**stata rubata.**
	STAH-tah roo-BAH-tah
a suitcase.	**una valigia.**
	OO-nah vah-LEE-jah
a briefcase.	**una valigetta.**
	OO-nah vah-lee-JEHT-tah
a carry-on.	**un bagaglio a mano.**
	oon bah-GAHL-lyoh ah
	MAH-noh
a suit bag.	**una borsa portabiti.**
	OO-nah BOHR-sah
	pohr-TAH-bee-tee
a trunk.	**un baule.**
	oon bah-OO-leh
golf clubs.	**mazze da golf.**
	MAHT-seh dah golf

For colors terms, see English / Italian Dictionary.

hard.	**rigido -a.**
	REE-jee-doh -ah
made out of _____	**fatto -a di _____**
	FAHT-toh -ah dee
canvas.	**tela.**
	TEH-lah
vinyl.	**vinile.**
	vee-NEE-leh
leather.	**pelle.**
	PEHL-leh
hard plastic.	**plastica dura.**
	PLAHS-tee-kah DOO-rah
aluminum.	**alluminio.**
	ahl-loo-MEE-nyoh

GETTING THERE

RENTING A VEHICLE

Is there a car rental
agency in the airport?

**C'è un'agenzia di autonoleggio in
aeroporto?**
*ch-EH oon-ah-jehn-TSEE-ah dee
ow-toh-noh-LEHD-joh een
ah-EH-roh-POHR-toh*

I have a reservation.

Ho una prenotazione.
*OH OO-nah preh-noh-tah-
TSYOH-neh*

Vehicle Preferences
I would like to rent _____

Vorrei noleggiare _____
vohr-RAY noh-lehd-JAH-reh

an economy car.

un'auto economica.
*oon-OW-toh eh-koh-NOH-
mee-kah*

a midsize car.

un'auto di media dimensione.
*oon-OW-toh dee MEH-dyah
dee-mehn-SYOH-neh*

a convertible.

una cabriolet.
OO-nah kah-bryoh-LEH

a van.

un furgoncino.
oon foor-gohn-CHEE-noh

a sports car.

un'auto sportiva.
oon-OW-toh spohr-TEE-vah

a 4-wheel-drive vehicle.

una quattro per quattro.
*OO-nah KWAHT-troh pehr
KWAHT-troh*

a motorcycle.

una moto.
OO-nah MOH-toh

a scooter.

uno scooter.
OO-noh scooter

Do you have one with ____	**C'è un'auto con ____** *ch-EH oon-OW-toh kohn*
air conditioning?	**climatizzatore?** *klee-mah-teed-zsah-TOH-reh*
a sunroof?	**tettuccio apribile?** *teht-TOOT-choh ah-PREE-bee-leh*
a CD player?	**lettore di CD?** *leht-TOH-reh dee chee-DEE*
an iPod connection?	**una porta per l'Ipod?** *OO-nah POHR-tah pehr LAY-pohd*
a GPS system?	**sistema di navigazione satellitare?** *sees-TEH-mah dee nah-vee-gah-TZYOH-neh sah-tehl-lee-TAH-reh*
a DVD player?	**lettore di DVD?** *leht-TOH-reh dee dee-voo-DEE*
child seats?	**seggiolini?** *sehd-djyoh-LEE-nee*
Do you have a ____	**C'è ____** *ch-EH*
smaller car?	**un'auto più piccola?** *oon-OW-toh PYOO PEEK-koh-lah*
bigger car?	**un'auto più grande?** *oon-OW-toh PYOO GRAHN-deh*
cheaper car?	**un'auto più economica?** *oon-OW-toh PYOO eh-koh-NOH-mee-kah*
Do you have a non-smoking car?	**C'è un'auto per non fumatori?** *ch-EH oon-OW-toh pehr nohn foo-mah-TOH-ree*
I need an automatic transmission.	**Mi serve un'auto con cambio automatico.** *mee SEHR-veh oon-OW-toh kohn KAHM-byoh ow-toh-MAH-tee-koh*

A standard transmission is okay.

Con cambio manuale va bene.
kohn KAHM-byoh mah-NWAH-leh vah BEH-neh

May I have an upgrade?

Posso avere una categoria superiore?
POHS-soh ah-VEH-reh OO-nah kah-teh-goh-REE-ah soo-peh-RYOH-reh

Money Matters

What's the daily / weekly / monthly rate?

Qual è la tariffa giornaliera / settimanale / mensile?
kwah-LEH lah tah-REEF-fah johr-nah-LYEH-rah / seht-tee-mah-NAH-leh / mehn-SEE-leh

What is the mileage rate?

Qual è la tariffa chilometrica?
kwah-LEH lah tah-REEF-fah kee-loh-MEH-tree-kah

How much is insurance?

Quanto costa l'assicurazione?
KWAHN-toh KOHS-tah lahs-see-koo-raht-SYOH-neh

Are there other fees?

Ci sono altri costi?
chee SOH-noh AHL-tree KOHS-tee

Is there a weekend rate?

C'è una tariffa per il weekend?
ch-EH OO-nah tah-REEF-fah pehr eel weekend

Technical Questions

What kind of gas does it take?

Che tipo di benzina bisogna mettere?
keh TEE-poh dee behn-DZEE-nah bee-ZOHN-njah MEHT-teh-reh

Do you have the manual in English?

Ha il manuale in inglese?
AH eel mah-NWA-leh een een-GLEH-zeh

Do you have an English booklet with the local traffic laws?	**Ha un codice della strada in inglese?** *AH oon KOH-dee-cheh DEHL-lah STRAH-dah een een-GLEH-zeh*

Car Troubles

The _____ doesn't work.

_____ non funziona. *nohn foon-TSYOH-nah*

See diagram on p54 for car parts.

It is already dented.	**È già ammaccata.** *EH JAH ahm-mahk-KAH-tah*
It is scratched.	**È graffiata.** *EH grahf-FYAH-tah*
The windshield is cracked.	**Il parabrezza è incrinato.** *eel pah-rah-BREHT-sah EH een-kree-NAH-toh*
The tires look low.	**Le gomme sembrano sgonfie.** *leh GOHM-meh SEHM-brah-noh SGOHN-fyeh*
It has a flat tire.	**Ha una gomma a terra.** *AH OO-nah GOHM-mah ah TEHR-rah*
Whom do I call for service?	**Chi chiamo per l'assistenza?** *kee KYAH-moh pehr lahs-sees-TEHN-tsa*
It won't start.	**Non parte.** *nohn PAHR-teh*
It's out of gas.	**È a secco.** *EH ah SEHK-koh*
The Check Engine light is on.	**La spia del motore è accesa.** *lah SPEE-ah dehl moh-TOH-reh EH aht-CHEH-zah*
The oil light is on.	**La spia dell'olio è accesa.** *lah SPEE-ah dehl-LOH-lyoh EH aht-CHEH-zah*
The brake light is on.	**La spia dei freni è accesa.** *lah SPEE-ah day FREH-nee EH aht-CHEH-zah*

GETTING THERE

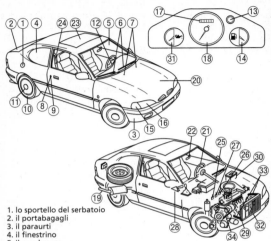

1. lo sportello del serbatoio
2. il portabagagli
3. il paraurti
4. il finestrino
5. il parabrezza
6. i tergicristalli
7. il liquido tergicristalli
8. la serratura
9. la serratura automatica
10. i pneumatici
11. le ruote
12. l'accensione
13. la spia
14. l'indicatore di livello di carburante
15. gli indicatori di direzione
16. i fanali
17. il contachilometri
18. il tachimetro
19. la marmitta
20. il cofano
21. il volante
22. lo specchietto retrovisore
23. il tettuccio apribile
24. la cintura di sicurezza

25. l'acceleratore
26. la frizione
27. il freno
28. il freno d'emergenza
29. il motore
30. la batteria
31. l'indicatore di livello dell'olio
32. il radiatore
33. il tubo del radiatore
34. la cinghia del ventilatore

It runs rough.	**Non va bene, fa rumore.** *nohn vah BEH-neh fah roo-MOH-reh*
The car is over-heating.	**L'auto si surriscalda.** *LOW-toh see soor-ree-SKAHL-dah*

Asking for Directions

Excuse me.	**Mi scusi.** *mee SKOO-zee*
How do I get to _____?	**Come si arriva a _____?** *KOH-meh see ahr-REE-vah ah*
Go straight.	**Vada dritto.** *VAH-dah DREET-toh*
Turn left.	**Giri a sinistra.** *JEE-ree ah see-NEES-trah*
Continue right.	**Continui a destra.** *kohn-TEE-nwee ah DEHS-trah*
It's on the right.	**E' sulla destra.** *EH SOOL-lah DEHS-trah*
Can you show me on the map?	**Può mostrarmi sulla cartina?** *PWOH mohs-TRAHR-mee SOOL-lah kahr-TEE-nah*
What are the GPS coordinates?	**Quali sono le coordinate satellitari?** *KWAH-lee SOH-noh leh koh-ohr-dee-NAH-teh sah-tehl-leh-TAH-ree*
How far is it from here?	**Quanto dista da qui?** *KWAHN-toh DEES-tah dah kwee*
Is this the right road for _____?	**Questa è la strada giusta per _____?** *KWEHS-tah EH lah STRAH-dah JOOS-tah pehr*
I've lost my way.	**Mi sono perso -a.** *mee SOH-noh PEHR-soh -sah*
Would you repeat that, please?	**Può ripetere, per favore?** *PWOH ree-PEH-teh-reh pehr fah-VOH-reh*
Thanks for your help.	**Grazie per l'aiuto.** *GRAH-tsyeh pehr lah-YOO-toh*

GETTING THERE

For full coverage of direction-related terms, see p5.

Sorry, Officer

What is the speed limit?	**Qual è il limite di velocità?** *kwah-LEH eel LEE-mee-teh dee* *veh-loh-chee–TAH*
I wasn't going that fast.	**Non andavo così veloce.** *nohn ahn-DAH-voh koh-ZEE* *veh-LOH-cheh*
How much is the fine?	**Quant'è la multa?** *kwahn-TEH lah MOOL-tah*
Where do I pay the fine?	**Dove si paga la multa?** *DOH-veh see PAH-gah lah MOOL-tah*
Do I have to go to court?	**Devo andare in tribunale?** *DEH-voh ahn-DAH-reh een tree-* *boo-NAH-leh*
I had an accident.	**Ho avuto un incidente.** *OH ah-VOO-toh oon een-chee-* *DEHN-teh*
The other driver hit me.	**L'altro autista mi ha investito -a.** *LAHL-troh ow-TEES-tah mee AH* *een-vehs-TEE-toh -tah*

Road Signs

Limite di velocità	Speed Limit
Stop	Stop
Dare la precedenza	Yield
Pericolo	Danger
Strada senza sbocco	No Exit
Senso unico	One Way
Vietato l'accesso	Do Not Enter
Strada chiusa	Road Closed
Pagamento pedaggio	Toll
Solo contanti	Cash Only
Parcheggio vietato	No Parking
Tariffa di parcheggio	Parking fee
Parcheggio	Parking garage
ZTL (zona a traffico limitato)	Restricted traffic zone

I'm at fault.

È colpa mia.
EH KOHL-pah MEE-ah

BY TAXI

Where is the taxi stand?

Dov'è la fermata dei taxi?
doh-VEH lah fehr-MAH-tah day taxi

Is there a limo / bus / van
for my hotel?

**C'è un servizio di limousine / bus /
navetta per il mio hotel?**
*ch-EH oon sehr-VEE-tsyoh dee
limousine / boos / nah-VEHT-tah
pehr eel MEE-oh hotel*

I need to get to _____.

Devo andare a _____.
DEH-voh ahn-DAH-reh ah

How much will that cost?

Quanto mi costa?
KWAHN-toh mee KOHS-tah

How long will it take?

Quanto tempo ci vuole?
KWAHN-toh TEHM-poh chee VWOH-leh

Can you take me / us to the
train / bus station?

**Può portarmi / portarci alla
stazione dei treni / degli autobus?**
*PWOH pohr-TAHR-mee / chee
AHL-lah stah-TSYOH-neh day
TREH-nee / DEHL-lye OW-toh-boos*

GETTING THERE

Listen Up: Taxi Lingo

Salga! *SAHL-gah*	Get in!
Lasci i bagagli, faccio io. *LAH-shee ee bah-GAHL-lyee FAH-choh EE-oh*	Leave your luggage, I got it.
Sono cinque euro al pezzo. *SOH-noh CHEEN-kweh ehu- roh AHL PEH-tsoh*	It's 5 Euros for each bag.
Quanti passeggeri? *KWAHN-tee pahs-sed-JEH-ree*	How many passengers?
Ha fretta? *AH FREHT-tah*	Are you in a hurry?

I am in a hurry.	**Ho fretta.** *OH FREHT-tah*
Slow down.	**Rallenti.** *rahl-LEHN-tee*
Am I close enough to walk?	**Sono abbastanza vicino da andarci a piedi?** *SOH-noh ahb-bahs-TAHN-tsah vee-CHEE-noh dah ahn-DAHR-chee ah PYEH-dee*
Let me out here.	**Mi lasci qui.** *mee LAH-shee kwee*
That's not the correct change.	**Il resto non è giusto.** *eel REH-stoh nohn EH JOOS-toh*

BY TRAIN

How do I get to the train station?	**Come si arriva alla stazione ferroviaria?** *KOH-meh see ahr-REE-vah AHL-lah stah-TSYOH-neh fehr-roh-VYAH-ryah*
Would you take me to the train station?	**Può portarmi alla stazione ferroviaria?** *PWOH pohr-TAHR-mee AHL-lah stah-TSYOH-neh fehr-roh-VYAH-ryah*
How long is the trip to ____?	**Quanto ci vuole fino a ____?** *KWAHN-toh chee VWOH-leh FEE-noh ah*
When is the next train?	**Quand'è il prossimo treno?** *kwahn-DEH eel PROHS-see-moh TREH-noh*
Is there an intercity / regional / high-speed train for my route?	**C'è un treno intercity / regionale / ad alta velocità per il mio percorso?** *CHEH oon TREH-noh intercity / reh-jyoh-NAH-leh / ahd AHL-tah veh-loh-CEE-tah pehr eel MEE-oh pehr-KOHR-soh*
Do you have a schedule / timetable?	**Ha un orario?** *AH oon oh-RAH-ryoh*
Do I have to change trains?	**Devo cambiare treni?** *DEH-voh kahm-BYAH-reh TREH-nee*

Where is the machine to validate my rail ticket?	**Dov'è la macchinetta per timbrare il biglietto?** *doh-VEH lah mahk-kee-NEHT-tah pehr teem-BRAH-reh eel bee-JYEHT-toh*
a one-way ticket	**un biglietto di sola andata** *oon beel-LYEHT-toh dee SOH-lah ahn-DAH-tah*
a round-trip ticket	**un biglietto di andata e ritorno** *oon beel-LYEHT-toh dee ahn-DAH-tah eh ree-TOHR-noh*
Which platform does it leave from?	**Da che binario parte?** *dah keh bee-NAH-ryoh PAHR-teh*
Is there a bar car?	**C'è una carrozza bar?** *ch-EH OO-nah kahr-ROHT-sah bar*
Is there a dining car?	**C'è una carrozza ristorante?** *ch-EH OO-nah kahr-ROHT-sah ree-stoh-RAHN-teh*
Which car is my seat in?	**In quale carrozza è il mio posto?** *een KWAH-leh kahr-ROHT-sah EH eel MEE-oh POHS-toh*
Can I plug in my laptop?	**Posso collegare il mio portatile?** *POHS-soh kohl-leh-GAH-reh eel MEE-oh pohr-tah-TEE-leh*
Is this seat taken?	**È occupato questo posto?** *EH ohk-koo-PAH-toh KWEHS-toh POHS-toh*
Where is the next stop?	**Dov'è la prossima fermata?** *doh-VEH lah PROHS-see-mah fehr-MAH-tah*
How many stops to ____?	**Quante fermate fino a ____?** *KWAHN-teh fehr-MAH-teh FEE-noh ah*
What's the train number and destination?	**Qual è il numero del treno e la destinazione?** *kwah-LEH eel NOO-meh-roh dehl TREH-noh eh lah dehs-tee-nah-TSYOH-neh*

BY BUS

How do I get to the bus station?

Come si arriva alla stazione degli autobus?
KOH-meh see ahr-REE-vah ahl-LAH stah-TSYOH-neh DEHL-lye OW-toh-boos

Would you take me to the bus station?

Può portarmi alla stazione degli autobus?
PWOH pohr-TAHR-mee ahl-LAH stah-TSYOH-neh DEHL-lye OW-toh-boos

May I have a bus schedule?

Posso avere un orario degli autobus?
POHS-soh ah-VEH-reh oon oh-RAH-ryoh DEHL-lye OW-toh-boos

Which bus goes to _____?

Quale autobus va a _____?
KWAH-leh OW-toh-boos vah ah

Where does it leave from?

Da dove parte?
dah DOH-veh PAHR-teh

How long does the bus take?

Quanto tempo ci impiega l'autobus?
KWAHN-toh TEHM-poh chee eem-PYEH-gah LOW-toh-boos

How much is it?

Quanto costa?
KWAHN-toh KOHS-tah

Is there an express bus?

C'è un autobus veloce?
ch-EH oon OW-toh-boos veh-LOH-cheh

Does it make local stops?	**Fa fermate locali?**
	fah fehr-MAH-teh loh-KAH-lee
Does it run at night?	**Fa servizio notturno?**
	fah sehr-VEE-tsyoh noht-TOOR-noh
When is the next bus?	**Quand'è il prossimo autobus?**
	kwahn-DEH eel PROHS-see-moh OW-toh-boos
a one-way ticket	**un biglietto di sola andata**
	oon beel-LYEHT-toh dee SOH-lah ahn-DAH-tah
a round-trip ticket	**un biglietto di andata e ritorno**
	oon beel-LYEHT-toh dee ahn-DAH-tah eh ree-TOHR-noh
How long will the bus be stopped?	**Per quanto tempo si ferma l'autobus?**
	Pehr KWAHN-toh TEHM-poh see FEHR-mah LOW-toh-boos
Is there an air conditioned bus?	**C'è un autobus con aria condizionata?**
	ch-EH oon OW-toh-boos kohn AH-ryah kohn-dee-tsyoh-NAH-tah
Is this seat taken?	**È occupato questo posto?**
	EH ohk-koo-PAH-toh KWEHS-toh POHS-toh

GETTING THERE

Ticket etiquette

Travelers in Italy often need to validate their tickets before embarking. In a train station, search for a yellow **macchina obliteratrice** (stamping machine), which prints the time and date in the space marked **convalida** (validation). These usually stand at the end of each **binario** (platform). Many subways have a similar system, but buses validate onboard. If you forget—or can't reach the machine in a crowd—write the details in pen.

Where is the next stop?
Dov'è la prossima fermata?
doh-VEH lah PROHS-see-mah fehr-MAH-tah

Please tell me when we reach ____.
Mi dice quando arriviamo a ____, per favore.
mee DEE-cheh KWAHN-doh ahr-ree-VYAH-moh ah pehr fah-VOH-reh

Let me off here.
Mi lasci qui.
mee LAH-shee kwee

BY BOAT OR SHIP

Would you take me to the port?
Può portarmi al porto?
PWOH pohr-TAHR-mee ahl POHR-toh

When does the ship sail?
Quando salpa la nave?
KWAHN-doh SAHL-pah lah NAH-veh

How long is the trip?
Quanto dura il viaggio?
KWAHN-toh DOO-rah eel VYAHD-joh

Where are the life preservers?
Dove sono i salvagenti?
DOH-veh SOH-noh ee sahl-vah-JEHN-tee

I would like a private cabin.
Vorrei una cabina privata.
vohr-RAY OO-nah kah-BEE-nah pree-VAH-tah

Is the trip rough?
È una traversata agitata?
EH OO-nah trah-vehr-SAH-tah ah-jee-TAH-tah

I feel seasick.
Ho mal di mare.
OH eel mahl dee MAH-reh

I need some seasick pills.
Ho bisogno di pastiglie antinausea.
OH bee-ZOHN-nyoh dee pahs-TEEL-lyeh ahn-tee-NOW-seh-ah

Where is the bathroom?
Dov'è la toilette?
doh-VEH lah twah-LEHT

Does the ship have a casino?	**C'è un casinò sulla nave?** *ch-EH oon kah-zee-NOH SOOL-lah NAH-veh*
Will the ship stop at ports along the way?	**La nave si ferma nei porti lungo il tragitto?** *lah NAH-veh seeFEHR-mah NAY POHR-tee LOON-goh eel trah-JEET-oh*

BY SUBWAY

Where's the subway station?	**Dov'è la stazione della metropolitana?** *doh-VEH lah stah-TSYOH-neh DEHL-lah MEH-troh-poh-lee-ta-nah*
Where can I buy a ticket?	**Dove si comprano i biglietti?** *DOH-veh see KOHM-prah-noh ee beel-LYEHT-tee*

SUBWAY TICKETS

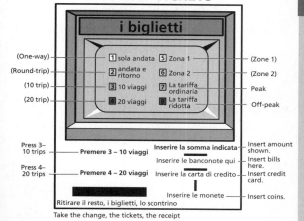

	i biglietti	
(One-way)	**1** sola andata	**5** Zona 1	(Zone 1)
(Round-trip)	**2** andata e ritorno	**6** Zona 2	(Zone 2)
(10 trip)	**3** 10 viaggi	**7** La tariffa ordinaria	Peak
(20 trip)	**4** 20 viaggi	**8** La tariffa ridotta	Off-peak

Press 3– 10 trips — **Premere 3 – 10 viaggi** **Inserire la somma indicata** — Insert amount shown.

Inserire le banconote qui — Insert bills here.

Press 4– 20 trips — **Premere 4 – 20 viaggi** Inserire la carta di credito — Insert credit card.

Inserire le monete — Insert coins.

Ritirare il resto, i biglietti, lo scontrino

Take the change, the tickets, the receipt

Could I have a map of the subway?	**Potrei avere una cartina del metrò?** *poh-TREH-ee ah-VEH-reh OO-nah kahr-TEE-nah dehl meh-TROH*
Which line for ____?	**Che linea devo prendere per ____?** *keh LEE-neh-ah DEH-voh PREHN-deh-reh pehr*
Is this the right line for ____?	**Questa è la linea giusta per ____?** *KWEHS-tah EH lah LEE-neh-ah JOOS-tah pehr*
Which stop is it for ____?	**Qual è la fermata per ____?** *kwah-LEH lah fehr-MAH-tah pehr*
How many stops is it to ____?	**Quante fermate per ____?** *KWAHN-teh fehr-MAH-teh pehr*
Is the next stop ____?	**La prossima fermata è ____?** *lah PROHS-see-mah fehr-MAH-tah EH*
Where are we?	**Dove siamo?** *DOH-veh see-AH-moh*
Where do I change to ____?	**Dove devo cambiare per ____?** *DOH-veh DEH-voh kahm-BYAH-reh pehr*
What time is the last train to ____?	**A che ora parte l'ultimo treno per ____?** *ah keh OH-rah PAHR-teh LOOL-tee-moh TREH-noh pehr*

CONSIDERATIONS FOR TRAVELERS WITH SPECIAL NEEDS

Do you have wheelchair access?	**C'è accesso per la sedia a rotelle?** *ch-EH atch-CHESS-oh pehr lah SEH-dyah ah roh-TEHL-leh*
Do you have elevators? Where?	**Ci sono ascensori? Dove?** *chee SOH-noh ah-shehn-SOH-ree DOH-veh*

Do you have ramps?
Where?
Are the restrooms
wheelchair accessible?

Ci sono rampe? Dove?
chee SOH-noh RAHM-peh DOH-veh
**Le toilette hanno accesso per la
sedia a rotelle?**
*leh twa-LEHT AHN-noh atch-
CHESS-oh pehr lah SEH-dyah ah
roh-TEHL-leh*

Do you have audio
assistance for the hearing
impaired?
I am deaf.

C'è assistenza audio per ipoudenti?
*ch-EH ahs-sees-TEHN-tsa OW-dyoh
pehr EE-poh-oo-DEHN-tee*
Sono ipoudente.
SOH-noh ee-poh-oo-DEHN-teh

May I bring my service dog?

Posso portare il mio cane guida?
*POHS-soh pohr-TAH-reh eel MEE-
oh KAH-neh GWEE-dah*

I am blind.

Sono ipovedente.
SOH-noh ee-poh-veh-DEHN-teh

I need to charge my
power chair.

**Devo caricare la mia sedia a
rotelle motorizzata.**
*DEH-voh kah-ree-KAH-reh lah
MEE-ah SEH-dyah ah roh-TEHL-leh
moh-toh-reez-ZAH-tah*

CHAPTER THREE

LODGING

This chapter will help you find the right accommodations, at the right price—and the amenities you might need during your stay.

ROOM PREFERENCES

Please recommend _____

Per favore, mi consigli _____
pehr fah-VOH-reh mee kohn-SEEL-lyee

a clean hostel.

un ostello pulito.
oon oss-TELL-loh poo-LEE-toh

a moderately priced hotel.

un albergo non caro.
oon ahl-BEHR-goh nohn KAH-roh

a moderately priced B&B.

una pensione non cara.
OO-nah pehn-SYOH-neh nohn KAH-rah

a good hotel / motel.

un buon hotel / motel.
oon BWON hotel / motel

Does the hotel have _____

L'hotel ha _____
loh-TEL AH

an indoor /outdoor pool?

una piscina coperta / all'aperto?
OO-nah pee-SHEEH-nah koh-PEHR-tah/ahl-lah-PEHR-toh

suites?

delle suite?
DEHL-leh suite

a fitness center?

una palestra?
OO-nah pah-LEHS-trah

a spa?

un centro benessere?
CHEN-troh beh-NEHS-seh-reh

a private beach?

una spiaggia privata?
OO-nah SPYAD-jah pree-VAH-tah

a tennis court?

un campo da tennis?
oon KAHM-poh dah tennis

air conditioned rooms?

stanze con aria condizionata?

free Wi-Fi?

STAHN-tseh kohn AH-ryah kohn-dee-tsyo-NAH-tah
Wi-Fi gratuito?
Wi-Fi grah-TOOH-toh

I would like a room for ____.

Vorrei una stanza per ____.
vohr-RAY OO-nah STAHN-tsah pehr

For full coverage of numbers, see p7.

I would like ____

Vorrei ____
vohr-RAY

a king-sized bed.

un letto matrimoniale king size.
oon LET-toh mah-tree-moh-NYAH-leh (king size)

a double bed.

un letto matrimoniale.
oon LET-toh mah-tree-moh-NYAH-leh

Listen Up: Reservations Lingo

Non abbiamo stanze libere. *nohn ahb-BYAH-moh STAHN-tseh LEE-beh-reh*	We have no vacancies.
Quanto si ferma? *KWAHN-toh see FEHR-mah*	How long will you be staying?
Per fumatori o non fumatori? *pehr foo-mah-TOH-ree oh nohn foo-mah-TOH-ree*	Smoking or non smoking?
Abbiamo solo mezza pensione / pensione completa. *ahb-BYAH-moh SOH-loh meh-dzah pehn-SYOH-neh / pehn-SYOH-neh kohm-PLEH-tah*	We only have full / half board.
Devo tenere il passaporto fino a domani. *DEH-voh teh-NEH-reh eel pahs-sah-POHR-toh FEE-noh ah doh-MAH-nee*	I need to keep your passport overnight.

twin beds.	**due letti singoli.**
	DOO-eh LET-tee SEEN-goh-lee
adjoining rooms.	**stanze adiacenti.**
	STAHN-tseh ah-dyah-CHEN-tee
a smoking room.	**una stanza per fumatori.**
	OO-nah STAHN-tsah pehr
	foo-mah-TOH-ree
a nonsmoking room.	**una stanza per non fumatori.**
	OO-nah STAHN-tsah pehr nohn
	foo-mah-TOH-ree
a private bathroom.	**un bagno privato.**
	oon BAHN-nyoh pree-VAH-toh
a shower.	**una doccia.**
	OO-nah DOT-chah
a bathtub.	**una vasca da bagno.**
	OO-nah VAHS-kah dah BAHN-nyoh
air conditioning.	**l'aria condizionata.**
	LAH-ryah kohn-dee-tsyoh-
	NAH-tah
television.	**la televisione.**
	lah teh-leh-vee-ZYOH-neh
cable.	**la televisione via cavo.**
	lah teh-leh-vee-ZYOH-neh
	VEE-ah KAH-voh
satellite TV.	**la TV satellitare.**
	lah tee-VOO
	sah-tehl-lee-TAH-reh
a telephone.	**un telefono.**
	oon teh-LEH-foh-noh
Internet access.	**accesso a Internet.**
	atch-CHESS-oh ah internet
Wi-Fi.	**Wi-Fi**
	Wi-Fi
a refrigerator.	**un frigorifero.**
	oon free-goh-REE-feh-roh
a beach view.	**vista sulla spiaggia.**
	VEES-tah SOOL-lah SPYAD-jah

a city view.	**vista sulla città.**
	VEES-tah SOOL-lah ceet-TAH
a kitchenette.	**un angolo cottura.**
	oon AHN-goh-loh koht-TOO-rah
a balcony.	**un balcone.**
	oon bahl-KOH-neh
a suite.	**una suite.**
	OO-nah suite
a penthouse.	**un attico.**
	oon AHT-tee-koh
I would like a room ____	**Vorrei una stanza ____**
	vohr-RAY OO-nah STAHN-tsah
on the ground floor.	**a pianterreno.**
	ah pyahn-tehr-REH-noh
near the elevator.	**vicino all'ascensore.**
	vee-CHEE-noh ahl-ash-ehn-SOH-reh
near the stairs.	**vicino alle scale.**
	vee-CHEE-noh AHL-leh SKAH-leh
near the pool.	**vicino alla piscina.**
	vee-CHEE-noh AHL-lah pee-SHEE-nah
away from the street.	**lontano dalla strada.**
	lohn-TAH-noh DAHL-lah STRAH-dah
I would like a corner room.	**Vorrei una stanza d'angolo.**
	vohr-RAY OO-nah STAHN-tsah DAHN-goh-loh
Do you have ____	**C'è ____**
	ch-EH
a crib?	**una culla?**
	OO-nah KOOL-lah
a foldout bed?	**un lettino pieghevole?**
	oon leht-TEE-noh pyeh-GHE-voh-leh

FOR GUESTS WITH SPECIAL NEEDS

I need a room with _____	**Mi serve una stanza con _____** *mee SEHR-veh OO-nah STAHN-tsah kohn*
wheelchair access.	**accesso per sedia a rotelle.** *atch-CHESS-oh pehr SEH-dyah ah roh-TEHL-leh*
services for the visually impaired.	**servizi per ipovedenti.** *sehr-VEET-see pehr ee-poh-veh-DEHN-tee*
services for the hearing impaired.	**servizi per ipoudenti.** *sehr-VEET-see pehr ee-poh-oo-DEHN-tee*
I am traveling with a service dog.	**Viaggio con un cane guida.** *VYAHD-joh kohn oon KAH-neh GWY-dah*

MONEY MATTERS

I would like to make a reservation.	**Vorrei fare una prenotazione.** *vohr-RAY FAH-reh OO-nah preh-noh-tah-TSYOH-neh*
How much per night?	**Quanto costa a notte?** *KWAHN-toh KOHS-tah ah NOHT-teh*
Is breakfast included?	**È compresa la colazione?** *EH kohm-PREH-zah lah koh-lah-TSYOH-neh*
Do you have a _____	**C'è una tariffa _____** *ch-EH OO-nah tah-REEF-fah*
weekly / monthly rate?	**settimanale / mensile?** *seht-tee-mah-NAH-leh / mehn-SEE-leh*
a weekend rate?	**per il weekend?** *pehr eel weekend*
a special rate?	**una tariffa speciale** *OO-nah tah-REEF-fah speh-CHAH-leh*
a discount?	**uno sconto?** *OO-noh SKOHN-toh*

We will be staying for ____ days / weeks.

Staremo per ____ giorni / settimane.
stah-REH-moh pehr ___ JOHR-nee / seht-tee-MAH-neh

For full coverage of number terms, see p7.

When is checkout time?

A che ora è il check-out?
ah keh OH-rah EH eel check-out

For full coverage of time-related terms, see p12.

Do you accept credit cards?

Accettate la carta di credito?
ah-tcheht-TAH-teh lah KAHR-tah dee KREH-dee-toh

May I see a room?

Posso vedere la stanza?
POHS-soh veh-DEH-reh lah STAHN-tsah

Is there a service charge?

C'è una tariffa per il servizio?
ch-EH OO-nah tah-REEF-fah pehr eel sehr-VEE-tsyoh

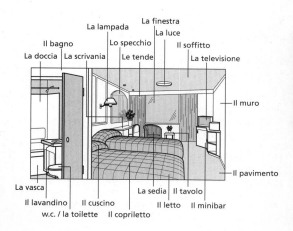

La lampada
La finestra
Il bagno
Lo specchio
La luce
La doccia   La scrivania
Le tende
Il soffitto
La televisione
Il muro
Il pavimento
La vasca
La sedia   Il tavolo
Il lavandino   Il cuscino
Il letto   Il minibar
w.c. / la toilette   Il copriletto

I'd like to speak with the manager.

Vorrei parlare con il direttore.
vohr-RAY pahr-LAH-reh kohn eel dee-reht-TOH-reh

IN-ROOM AMENITIES

I'd like _____

Vorrei _____
vohr-RAY

to place an international call.

fare una chiamata internazionale.
FAH-reh OO-nah KYAH-mah-tah een-tehr-nah-tsyoh-NAH-leh

to place a long-distance call.

fare una chiamata interurbana.
FAH-reh OO-nah KYAH-mah-tah een-tehr-oor-BAH-nah

directory assistance in English.

l'assistenza abbonati in inglese.
lahs-sees-TEHN-tsa ahb-boh-NAH-tee een een-GLEH-zeh

room service.

il servizio in camera.
eel sehr-VEE-tsyoh een KAH-meh-rah

Instructions for dialing the hotel phone

Per chiamare un'altra stanza, comporre il numero di stanza.
pehr kyah-MAH-reh oo-NAHL-trah STAHN-tsah kohm-POHR-reh eel NOO-meh-roh dee STAHN-tsah

To call another room, dial the room number.

Per chiamate locali, comporre prima il 9.
pehr kyah-MAH-teh loh-KAH-lee kohm-POHR-reh pree-MAH eel NOH-veh

To make a local call, dial 9 first.

Per chiamare il centralino, comporre lo 0.
pehr kyah-MAH-reh eel chen-trah-lee-noh kohm-POHR-reh loh DZEH-roh

To call the operator, dial 0.

maid service.

il servizio di pulizia.
*eel sehr-VEE-tsyoh dee
poo-lee-TSEE-ah*

the front desk operator.

il centralino alla reception.
*eel chen-trah-lee-noh
AHL-lah reception*

Do you have room service?

Offrite il servizio in camera?
*ohf-FREE-teh eel sehr-VEE-tsyoh
een KAH-meh-rah*

When is the kitchen open?

Quando apre la cucina?
*KWAHN-doh AH-preh lah koo-
CHEE-nah*

When is breakfast served?

A che ora servite la colazione?
*ah keh OH-rah sehr-VEE-teh
lah koh-lah-TSYOH-neh*

For time-related terms, see p12.

Do you offer massages?

Offrite il servizio massaggi?
*ohf-FREE-teh eel sehr-VEE-tsyoh
mahs-SAHD-jee*

Do you have a lounge?

Avete un salotto?
ah-VEH-teh oon sah-LOHT-toh

Do you have a business
center?

Avete un centro d'affari?
*ah-VEH-teh oon CHEN-troh
dahf-FAH-ree*

Do you serve breakfast?

Servite la colazione?
*sehr-VEE-teh lah koh-lah-
TSYOH-neh*

Do you have Wi-Fi in rooms,
or in the lobby only?

**Le stanze sono dotate di Wi-Fi o è
possibile utilizzarlo solo alla
reception?**
*Leh STAHN-tseh SOH-noh doh-TAH-
teh dee Wi-Fi oh EH pohs-SEE-bee-
leh uh-tee-LEE-tsar-loh SOH-loh
AHL-lah reh-SE-pchon*

May I have a newspaper
in the morning?

Posso avere il giornale al mattino?
*POHS-soh ah-VEH-reh eel
johr-NAH-leh ahl maht-TEE-noh*

Do you offer a laundry service?	**Offrite il servizio di lavanderia?** *ohf-FREE-teh eel sehr-VEE-tsyoh dee lah-vahn-deh-REE-ah*
Do you offer dry cleaning?	**Offrite il sevizio di lavasecco?** *ohf-FREE-the-eel-sehr-VEE-tsyoh dee lah-vah-SEHK-koh*
May we have _____	**Possiamo avere ___** *pohs-SYAH-moh ah-VEH-reh*
clean sheets today?	**lenzuola pulite oggi?** *lehn-TSWO-lah poo-LEE-tee OHD-jee*
more towels?	**altri asciugamani?** *AHL-tree ah-shoo-gah-MAH-nee*
more toilet paper?	**altra carta igienica?** *AHL-trah KAHR-tah ee-JEH-nee-kah*
extra pillows?	**altri cuscini?** *AHL-tree koo-SHEE-nee*
shampoo?	**dello shampoo?** *DEHL-loh SHAHM-poh*
toothpaste?	**del dentifricio?** *DEHL dehn-tee-FREE-tchoh*
a toothbrush?	**uno spazzolino da denti?** *OO-noh spah-tsoh-LEE-noh dah DEHN-tee*
an adapter?	**un adattatore?** *oon ah-daht-tah-TOH-reh*
a bottle opener?	**un apribottiglia?** *oon ah-pree-bot-TEEL-yah*
Do you have an ice machine?	**C'è un distributore di ghiaccio?** *ch-EH oon dees-tree-boo-TOH-reh dee GYAT-choh*

Did I receive any ____	**Ho ricevuto ____** *OH ree-cheh-VOO-toh*
messages?	**messaggi?** *mehs-SAHD-jee*
mail?	**posta?** *POHS-ta*
faxes?	**fax?** *fax*
A spare key, please.	**Una chiave di scorta, per favore.** *OO-nah KYAH-veh dee SKOHR-tah pehr fah-VOH-reh*
I'd like a wake up call.	**Vorrei il servizio di sveglia.** *vohr-RAY eel sehr-VEE-tsyoh dee ZVEHL-lyah*
For time-related terms, see p12. Do you have an alarm clock?	**C'è una sveglia?** *ch-EH OO-nah SVEHL-lyah*
Is there a safe in the room?	**C'è una cassaforte in camera?** *ch-EH OO-nah kahs-sah-FOHR-teh een KAH-meh-rah*
Does the room have a hair dryer?	**C'è un asciugacapelli in camera?** *ch-EH oon ah-SHOO-gah-kah-PEHL-lee een KAH-meh-rah*

For time-related terms, see p12.

LODGING

HOTEL ROOM TROUBLE

May I speak with the manager?	**Posso parlare con il direttore?** *POHS-soh pahr-LAH-reh kohn eel dee-REHT-toh-reh*
The ____ does not work.	**____ non funziona.** *nohn foon-TSYOH-nah*
television	**Il televisore** *eel teh-leh-vee-ZOH-reh*
telephone	**Il telefono** *eel teh-LEH-foh-noh*
air conditioning	**Il condizionatore d'aria** *eel kohn-dee-tsyoh-nah-TOH-reh DAH-ryah*
Internet access	**L'accesso a Internet** *laht-CHESS-oh ah internet*
cable TV	**La TV via cavo** *lah TV VEE-ah KAH-voh*
There is no hot water.	**Non c'è acqua calda.** *nohn ch-EH AHK-wah KAHL-dah*
The toilet is overflowing!	**L'acqua del water trabocca!** *LAHK-wah dehl VAH-ter trah-BOHK-kah*

This room is _____	**Questa stanza è _____** *KWEHS-tah STAHN-tsah EH*
too noisy.	**troppo rumorosa.** *TROHP-poh roo-moh-ROH-zah*
too cold.	**troppo fredda.** *TROHP-poh FREHD-dah*
too warm.	**troppo calda.** *TROHP-poh KAHL-dah*
dirty.	**sporca.** *SPOHR-kah*
This room has _____	**In questa stanza ci sono _____** *een KWEHS-tah STAHN-tsah chee SOH-noh*
bugs.	**degli insetti.** *DEHL-lyee een-SEHT-tee*
mice.	**dei topi.** *day TOH-pee*
I'd like a different room.	**Vorrei un'altra stanza.** *vohr-RAY oo-NAHL-trah STAHN-tsah*
Do you have a bigger room?	**C'è una stanza più grande?** *ch-EH OO-nah STAHN-tsah PYOO GRAHN-deh*
I locked myself out of my room.	**Mi sono chiuso fuori dalla stanza.** *Mee soh-noh kiw-soh fuoh-ree dahl-lah STAHN-tsah*
I lost my keys.	**Ho perso le chiavi.** *Oh PEHR-soh leh KYAH-vee*
Do you have a fan?	**C'è un ventilatore?** *ch-EH oon vehn-tee-lah-TOH-reh*
The sheets are not clean.	**Le lenzuola non sono pulite.** *leh lehn-TSWOH-lah nohn SOH-noh poo-LEE-teh*
The towels are not clean.	**Gli asciugamani non sono puliti.** *lyee ah-shoo-gah-MAH-nee nohn SOH-noh poo-LEE-tee*

LODGING

The room is not clean.	**La stanza non è pulita.** *lah STAHN-tsah nohn EH poo-LEE-tah*
The guests next door / above / below are being very loud.	**Gli ospiti a fianco / di sopra / di sotto fanno molto rumore.** *lyee OHS-pee-tee ah FYAHN-koh / dee SOH-prah / dee SOHT-toh FAHN-noh MOHL-toh roo-MOH-reh*

CHECKING OUT

I think this charge is a mistake.	**Credo questo addebito sia errato.** *KREH-doh KWEHS-toh ahd-DEH-bee-toh SEE-ah ehr-RAH-to*
Please explain this charge to me.	**Mi spieghi questo addebito, per favore.** *mee SPYEH-ghee KWEHS-toh ahd-DEH-bee-toh pehr fah-VOH-reh*
May I leave these bags?	**Posso lasciare queste borse?** *POHS-soh lah-SHAH-reh KWES-teh BOHR-tseh*
I'm missing _____.	**Non trovo_____ .** *Nohn TROH-voh*
I've lost _____.	**Ho perso_____.** *Oh PEHR-tsoh*
Thank you, we enjoyed our stay.	**Grazie, siamo stati contenti del soggiorno.** *GRAH-tsyeh see-AH-moh STAH-tee kohn-TEHN-tee dehl sohd-JOHR-noh*
The service was excellent.	**Il servizio è stato ottimo.** *eel sehr-VEE-tsyoh EH STAH-toh OHT-tee-moh*
The staff is very professional and courteous.	**Il personale è molto professionale e cortese.** *eel pehr-soh-NAH-leh EH MOHL-toh proh-fehs-syo-NAH-leh eh kohr-TEH-zeh*

Please call a cab for me.

Mi chiama un taxi, per favore?
*mee KYAH-mah oon taxi
pehr fah-VOH-reh*

Would someone please
get my bags?

**Qualcuno può prendere le mie
borse?**
*kwahl-KOO-noh PWOH PREHN-
deh-reh leh MEE-eh BOHR-seh*

HAPPY CAMPING

I'd like a site for _____

Vorrei un posto per_____
vohr-RAY oon POHS-toh pehr

  a tent.

   una tenda.
   OO-nah TEHN-dah

  a camper.

   un camper.
   oon KAHM-pehr

Are there _____

Ci sono_____
chee SOH-noh

  bathrooms?

   i bagni?
   ee BAHN-nyee

  showers?

   le docce?
   leh DOHT-cheh

Is there running water?

C'è acqua corrente?
ch-EH AHK-wah kohr-REHN-teh

Is the water drinkable?

L'acqua è potabile?
LAHK-wah EH poh-TAH-bee-leh

Where is the electrical
hookup?

Dove sono gli attacchi elettrici?
*DOH-veh SOH-noh lyee
aht-TAH-kee eh-LEHT-tree-keh*

CHAPTER FOUR

DINING

This chapter includes a menu reader and the language you need to communicate in a range of dining establishments and food markets.

FINDING A RESTAURANT

Would you recommend a good _____ restaurant?	**Può consigliarmi un buon ristorante _____** *PWOH kohn-seel-LYAHR-mee oon bwon ree-stoh-RAHN-teh*
local	**locale?** *loh-KAH-leh*
French	**francese?** *frahn-CHEH-zeh*
Chinese	**cinese?** *chee-NEH-zeh*
Japanese	**giapponese?** *jahp-poh-NEH-zeh*
Asian	**asiatico?** *ah-ZYAH-tee-koh*
steakhouse	**con specialità di carne?** *kohn speh-chah-lee-TAH dee KAHR-neh*
seafood	**con specialità di pesce?** *kohn speh-chah-lee-TAH dee PEH-sheh*

vegetarian	**vegetariano?**
	veh-jeh-tah-RYAH-noh
buffet-style	**a buffet?**
	ah buhf-FEH
budget	**economico?**
	eh-koh-NOH-mee-koh

Would you recommend a good pizza restaurant? **Può consigliarmi una buona pizzeria?**
PWOH kohn-seel-LYAHR-mee OO-nah bwon-nah peet-tseh REE-ah

Which is the best restaurant in town? **Qual è il miglior ristorante in città?**
kwah-LEH eel meel-LYOHR ree-stoh-RAHN-teh een cheet-TAH

Is there a late-night restaurant nearby? **C'è un ristorante qui vicino aperto fino a tardi?**
ch-EH oon ree-stoh-RAHN-teh kwee vee-CHEE-noh

Is there a restaurant that serves breakfast nearby? **C'è un ristorante qui vicino che serve la colazione?**
ch-EH oon ree-stoh-RAHN-teh kwee vee-CHEE- noh keh SEHR-veh lah koh-lah-TSYOH-neh

Is it very expensive? **È molto caro?**
EH MOHL-toh KAH-roh

Do I need a reservation? **Bisogna prenotare?**
bee-ZOHN-nyah preh-noh-TAH-reh

Do I have to dress up? **Bisogna vestirsi bene?**
bee-ZOHN-nyah vehs-TEER-see BEH-neh

Do they serve lunch? **Servono il pranzo?**
SEHR-voh-noh eel PRAHN-tsoh

What time do they open for dinner? **A che ora aprono per la cena?**
ah keh OH-rah AH-proh-noh pehr lah CHEH-nah

For lunch?	**Per il pranzo?** *pehr eel PRAHN-tsoh*
What time do they close?	**A che ora chiudono?** *ah keh OH-rah KYOO-doh-noh*
Do you have a take out menu?	**C'è un menu da asporto?** *ch-EH oon meh-NOO dah ahs-POHR-toh*
Do you have a bar?	**C'è un bar?** *ch-EH oon bar*
Is there a café nearby?	**C'è un caffè qui vicino?** *ch-EH oon kahf-FEH kwee vee-CHEE-noh*

GETTING SEATED

Are you still serving?	**Siete ancora aperti?** *SYEH-teh ahn-KOH-rah ah-PEHR-tee*
How long is the wait?	**Quanto c'è da aspettare?** *KWAHN-toh ch-EH dah ahs-peht-TAH-reh*
May I see a menu?	**Potrei vedere il menù?** *poh-TREH-ee veh-DEH-reh eel meh-NUH*
A table for ____, please.	**Un tavolo per ____, per favore.** *oon TAH-voh-loh pehr ____ pehr fah-VOH-reh*
For a full list of numbers, see p7.	
Do you have a quiet table?	**C'è un tavolo tranquillo?** *ch-EH oon TAH-voh-loh trahn-KWIL-loh*
Do you have highchairs?	**Avete seggioloni?** *Ah-VEH-teh seh-joh-LOH-nee*
May we sit outside / inside please?	**Possiamo sederci fuori / dentro, per favore?** *pohs-SYAH-moh seh-DEHR-chee FWO-ree / DEHN-troh pehr fah-VOH-reh*

Listen Up: Restaurant Lingo

È necessaria la giacca.
EH neh-chehs-SAH-ryah lah JAHK-kah

You'll need a jacket.

Mi dispiace, non sono permessi i pantaloni corti.
mee dee-SPYAH-cheh nohn SOH-noh pehr-MEHS-see ee pahn-tah-loh-nee KOHR-tee

I'm sorry, no shorts are allowed.

Posso portarle qualcosa da bere?
POHS-soh pohr-TAHR-leh kwahl-KOH-zah dah BEH-reh

May I bring you something to drink?

Gradisce la carta dei vini?
grah-DEESH-eh lah KAHR-tah day VEE-nee

Would you like to see a wine list?

Vuol sentire le nostre specialità?
vwol sehn-TEE-reh leh NOHS-treh speh-chah-lee-TAH

Would you like to hear our specials?

È pronto -a per ordinare?
EH PROHN-toh -ah pehr ohr-dee-NAH-reh

Are you ready to order?

Mi dispiace signore / signora, ma la sua carta di credito è stata rifiutata.
mee dee-SPYAH-cheh seen-NYOH-reh / seen-NYOH-rah mah lah SOO-ah KAHR-tah dee KREH-dee-toh EH STAH-tah ree-few-TAH-tah

I'm sorry sir / madame, your credit card was declined.

May we sit at the counter?	**Possiamo sederci al banco?**
	pohs-SYAH-moh seh-DEHR-chee
	ahl BAHN-koh
I'd like to order.	**Posso ordinare?**
	POHS-soh ohr-dee-NAH-reh

ORDERING

Do you have a special tonight?	**Qual è la specialità di questa sera?**
	kwah-LEH lah speh-chah-lee-TAH
	dee KWEHS-tah SEH-rah
What do you recommend?	**Cosa consiglia?**
	KOH-sah kohn-SEEL-lyah
May I see a wine list?	**Posso vedere la carta dei vini?**
	POHS-soh veh-DEH-reh lah KAHR-
	tah day VEE-nee
Do you serve wine by the glass?	**Servite il vino a bicchiere?**
	sehr-VEE-teh eel VEE-noh ah beek-
	KYEH-reh
May I see a drink list?	**Posso vedere la lista delle bevande?**
	POHS-soh veh-DEH-reh lah LEES-
	tah DEHL-leh beh-VAHN-de
I would like it cooked _____	**Lo / La vorrei _____**
	loh / lah vohr-RAY
rare.	**cotto -a al sangue.**
	KOHT-toh -ah ahl
	SAHN-gweh
medium rare.	**cotto -a quasi al sangue.**
	KOHT-toh -ah KWAH-zee ahl
	SAHN-gweh
medium.	**cotto -a mediamente.**
	KOHT-toh -ah meh-dyah-MEHN-teh
medium well.	**cotto -a abbastanza bene.**
	KOHT-toh -ah ahb-bah-STAHN-
	tsah BEH-ne
well.	**ben cotto -a.**
	behn KOHT-toh -ah

charred.	**rosolato -a.**
	roh-zoh-LAH-toh -ah
Do you have a ____ menu?	**Avete un menu____**
	ah-VEH-teh oon meh-NOO
children's	**per i bambini?**
	pehr ee bahm-BEE-nee
diabetic	**per diabetici?**
	pehr dyah-BEH-tee-chee
kosher	**kasher?**
	KAH-sher
gluten-free	**senza glutine?**
	SEHN-tsa GLUH-tee-neh
vegetarian	**vegetariano?**
	veh-jeh-tah-RYAH-noh
What is in this dish?	**Cosa c'è in questo piatto?**
	KOH-sah ch-EH een KWEHS-toh PYAHT-toh
How is it prepared?	**Come viene preparato?**
	KOH-meh VYEH-neh preh-pah-RAH-toh
What kind of oil is that cooked in?	**In che tipo di olio viene cotto?**
	een keh TEE-poh dee OH-lyoh VYEH-ne KOHT-toh
Do you have any low-salt dishes?	**Avete dei piatti a basso contenuto di sale?**
	ah BAHS-soh kohn-teh-NOO-toh dee SAH-leh
On the side, please.	**A parte, per favore.**
	ah PAHR-teh pehr fah-VOH-reh
May I make a substitution?	**Posso sostituire una cosa?**
	POHS-soh soh-stee-too-EE-reh OO-nah KOH-sah
I'd like to try that.	**Vorrei provare quello.**
	vohr-RAY proh-VAH-reh KWEHL-loh

DINING

Is that fresh?

È fresco -a?
EH FREHS-koh -ah

Waiter!

Cameriere -a!
kah-meh-RYEH-reh -ah

Extra butter, please.

Mi porta altro burro, per favore.
mee POHR-tah AHL-troh BOOR-roh pehr fah-VOH-reh

No butter, thanks.

Senza burro, grazie.
SEHN-tsa BOOR-roh GRAH-tsyeh

No cream, thanks.

Senza panna, grazie.
SEHN-tsa PAHN-nah GRAH-tsyeh

No salt, please.

Senza sale, per favore.
SEHN-tsa SAH-leh pehr-fah-VOH-reh

May I have some oil, please?

Mi porta un po' di olio, per favore?
mee POHR-tah oon POH dee oh-LYOH pehr fah-VOH-reh

More bread, please.

Altro pane, per favore.
AHL-troh PAH-neh pehr fah-VOH-reh

I am lactose intolerant.

Sono intollerante al lattosio.
SOH-noh een-tohl-leh-RAHN-teh ahl laht-TOH-zyoh

Would you recommend something without milk?

Mi consiglia qualcosa senza latte?
mee kohn-SEEL-lyah kwahl-KOH-zah SEHN-tsah LAHT-teh

I am allergic to____

Sono allergico -a ____
SOH-noh ahl-LEHR-jee-koh -kah

nuts.

a noci e nocciole.
ah NOH-chee eh noht-CHOH-leh

peanuts.

alle arachidi.
AHL-leh ah-RAH-kee-dee

seafood.

ai frutti di mare.
eye FROOT-tee dee MAH-reh

shellfish.	**a molluschi e crostacei.**
	ah mohl-LOOS-kee eh krohs-TAH-cheh-ee
Water, please?	**Acqua, per favore?**
	AHK-wah pehr fah-VOH-reh
still water	**acqua naturale**
	AHK-wah nah-too-RAH-leh
sparkling water	**acqua frizzante**
	AHK-wah free-DZAHN-te
with ice	**con ghiaccio**
	kohn GYAT-choh
without ice	**senza ghiaccio**
	SEHN-tsah GYAT-choh
I'm sorry, I don't think this is what I ordered.	**Scusi, ma non credo di aver ordinato questo.**
	SKOO-zee mah nohn KREH-doh dee ah-VEHR ohr-dee-NAH-toh KWEHS-toh
My meat is a little over / under cooked.	**La carne è un po' troppo cotta / non è abbastanza cotta.**
	lah KAHR-neh EH oon POH TROHP-poh KOHT-tah /nohn eh ahb-bah-STAHN-tsah KOHT-tah
My vegetables are a little over / under cooked.	**Le verdure sono un po' troppo cotte / non sono abbastanza cotte.**
	leh vehr-DOO-reh SOH-noh oon POH TROHP-poh KOHT-teh / nohn SOH-noh ahb-bah-STAHN-tsah KOHT-teh
There's a bug in my food!	**C'è un insetto nel mio cibo!**
	ch-EH oon een-SEHT-toh nehl MEE-oh CHEE-boh
May I have a another?	**Me ne porta un altro?**
	meh neh POHR-tah oon AHL-troh
A dessert menu, please.	**Il menu dei dolci, per favore.**
	eel MEH-noo day DOHL-chee pehr fah-VOH-reh

DINING

DRINKS

Carafes of **vino della casa** (house wine) are usually three sizes: **litro**
LEE-troh (liter), **mezzo litro** (half liter) and **un quarto** (quarter), akin
to two generous glasses. Remember that standard wine bottles hold
750ml; it's easy to underestimate the 25% extra wallop of a liter (and
Italians frown on public drunkenness, even while encouraging **un po'
di vino**—a little wine—with midday and evening meals).

alcoholic	**alcooliche**
	ahl-KOH-lee-keh
cocktail	**cocktail**
	KOK-teyl
neat / straight	**liscio -a**
	LEE-shoh -ah
on the rocks	**con ghiaccio**
	kohn GYAT-choh
with (seltzer or soda)	**con selz / acqua frizzante**
water	*kohn seltz / AHK-wah freet-TSAHN-teh*
beer	**birra**
	BEER-rah
wine	**vino**
	VEE-noh
house wine	**vino della casa**
	VEE-noh DEHL-lah KAH-zah
sweet wine	**vino dolce**
	VEE-noh DOHL-cheh
dry white wine	**vino bianco secco**
	VEE-noh BYAHN-koh SEHK-koh
rosé	**rosato**
	roh-ZAH-toh
light-bodied wine	**vino leggero**
	VEE-noh lehd-JEH-roh
red wine	**vino rosso**
	VEE-noh ROHS-soh

How Do You Take It?

Prendere generally means to take. But if a bartender asks, **Prende qualcosa?**, he's not inviting you to steal his fancy corkscrew. He's asking what you'd like to drink.

full-bodied wine	**vino corposo**
	VEE-noh kohr-POH-zoh
sparkling sweet wine	**spumante**
	spoo-MAHN-teh
Champagne	**Champagne**
	shahm-PAHN-nyeh
liqueur	**liquore**
	lee-KWOH-reh
brandy	**brandy**
	brandy
cognac	**cognac**
	cognac
gin	**gin**
	gin
vodka	**vodka**
	vodka
rum	**rum**
	room
non-alcoholic	**analcolici**
	ah-nahl-KOH-lee-chee
hot chocolate	**cioccolata calda**
	chohk-koh-LAH-tah KAHL-dah
lemonade	**limonata**
	lee-moh-NAH-tah
milk shake	**frappè**
	frahp-PEH
milk	**latte**
	LAHT-teh

DINING

tea	**tè** *teh*
coffee	**caffè** *kahf-FEH*
cappuccino	**cappuccino** *kahp-pooch-CHEE-noh*
espresso	**caffè** *kahf-FEH*
iced coffee	**caffè freddo** *kahf-FEH FREHD-doh*
concentrated	**ristretto** *rees-TREHT-toh*
with a drop of alcohol	**corretto** *kohr-REHT-toh*
with a drop of milk	**macchiato** *mahk-KYAH-toh*
fruit juice	**succo di frutta** *SOOK-koh dee FROOT-tah*

For a full list of fruits, see p104.

SETTLING UP

Check, please.	**Il conto, per favore.** *eel KOHN-toh pehr fah-VOH-reh*
I'm full!	**Sono sazio -a!** *SOH-noh SAH-tsyoh -tsyah*
The meal was excellent.	**Il cibo era squisito.** *eel CHEE-boh EH-rah skwee-ZEE-toh*
There's a problem with my bill.	**C'è un problema con il conto.** *ch-EH oon proh-BLEH-mah kohn eel KOHN-toh*
Is the tip included?	**La mancia è inclusa?** *lah MAHN-chah EH een-KLOO-zah*
My compliments to the chef!	**I miei complimenti al cuoco!** *ee MEE-eh-ee kohm-plee-MEHN-tee ahl KWO-koh*

MENU READER

Italian cuisine varies broadly from region to region, but we've tried to make our list of classic dishes as encompassing as possible.

APPETIZERS (ANTIPASTI)

affettati misti: mixed cold cuts and vegetables (often buffet)
ahf-feht-TAH-tee MEES-tee

prosciutto: ham
proh-SHOOT-toh

  prosciutto cotto: cooked ham
  proh-SHOOT-toh KOHT-toh

  prosciutto crudo: air cured ham
  proh-SHOOT-toh KROO-doh

bruschetta: toasted bread slices with various toppings
broos-KEHT-tah

crocchette: croquettes
krohk-KEHT-teh

crostini: little toasts topped with a variety of ingredients
krohs-TEE-nee

insalata di mare: seafood salad
een-sah-LAH-tah dee MAH-reh

mozzarella con pomodori: mozzarella and tomatoes
moht-tsah-REHL-lah kohn poh-moh-DOH-ree

insalata di nervetti: calf's foot and veal shank salad
een-sah-LAH-tah dee nehr-VEHT-tee

olive: olives
oh-LEE-veh

peperonata: mixed sweet peppers stewed with tomatoes
peh-peh-roh-NAH-tah

SALADS (INSALATE)

insalata caprese: tomato and mozzarella salad
een-sah-LAH-tah kah-PREH-zeh

insalata mista: mixed salad
een-sah-LAH-tah MEES-tah

insalata verde: green salad
een-sah-LAH-tah VEHR-deh

DINING

insalata di crescione: watercress salad
een-sah-LAH-tah dee kreh-SHOH-neh
insalata di lattuga romana: romaine salad
een-sah-LAH-tah dee laht-TOO-gah roh-MAH-nah
insalata di pomodori: tomato salad
een-sah-LAH-tah dee poh-moh-DOH-ree
insalata di rucola: arugula / rocket salad
een-sah-LAH-tah dee ROO-koh-lah
insalata di spinaci: spinach salad
een-sah-LAH-tah dee spee-NAH-chee

PIZZA TYPES

bianca: white, without tomato sauce
BYAHN-kah
capricciosa: with a mixture of toppings (artichoke hearts, ham, olives, pickled mushrooms, capers, sometimes egg)
kah-preet-CHOH-zah
con funghi: with mushrooms
kohn FOON-gy
marinara: with tomato sauce, garlic, capers, oregano, and sometimes anchovies
mah-ree-NAH-rah
margherita: with tomato sauce, basil, and mozzarella
mahr-ghe-REE-tah
napoletana: with fresh tomatoes, garlic, oregano, anchovies, and mozzarella; toppings may vary
nah-poh-leh-TAH-nah
quattro stagioni: (four seasons) same toppings as capricciosa
KWAHT-troh stah-JOH-nee
siciliana: usually with capers, onions, and anchovies
see-chee-LYAH-nah

SAUCES (SALSE)

bagna cauda: hot anchovy, garlic, and oil dip
BAHN-nyah KOW-dah
pizzaiola: tomato sauce with onion and garlic
PEET-sah-YOH-lah

salsa verde: green sauce with parsley, anchovies, capers, and garlic
SAHL-sah VERH-deh

pesto: basil, pine nuts, and garlic sauce
PEHS-toh

SOUPS (ZUPPE)

acquacotta: Tuscan vegetable soup with poached egg
AHK-wah KOHT-tah

minestrone: mixed vegetable soup
mee-nehs-TROH-neh

ribollita: Tuscan bean and bread vegetable soup
ree-bohl-LEE-tah

zuppa: soup
DZOOP-pah

di pane: bread (recipes vary greatly)
dee PAH-neh

fagioli e farro: beans and spelt
fah-JOH-lee eh FAHR-roh

porri e patate: leek and potato
POHR-ree eh pah-TAH-teh

pavese: bread and egg, Pavia style
pah-VEH-zeh

PASTA DISHES

agnolotti: meat-stuffed pasta
ahn-nyoh-LOHT-tee

bucatini: hollow thick spaghetti, usually all'amatriciana (with a pancetta and tomato sauce)
boo-kah-TEE-nee

cannelloni: stuffed pasta tubes, topped with a sauce and baked
kahn-nehl-LOH-nee

cappellacci alla ferrarese: pumpkin-stuffed pasta, Ferrara style
kahp-pehl-LAHT-chee AHL-lah fehr-rah-REH-zeh

cappelletti: meat-stuffed pasta usually served in broth
kahp-pehl-LEHT-tee

fusilli: spiral-shaped pasta
foo-ZEEL-lee

gnocchi: potato dumplings
NYOHK-kee

pansotti: swiss chard- and herbs-stuffed pasta, usually served with walnut sauce, a Ligurian specialty
pahn-SOHT-tee

pappardelle alla lepre: wide pasta ribbons with hare sauce
pahp-pahr-DEHL-leh AHL-lah LEH-preh

penne strascicate: pasta quills sauteed with meat sauce
PEHN-neh strah-shee-KAH-teh

PASTA SAUCES (SUGHI)

alfredo: butter and parmesan cheese
ahl-FREH-doh

amatriciana: pancetta and tomato
ah-mah-tree-CHAH-nah

arrabbiata: tomato and hot pepper
ahr-rahb-BYAH-tah

bolognese: ground meat and tomato
boh-lohn-NYEH-zeh

carbonara: pancetta and egg
kahr-boh-NAH-rah

panna: cream
PAHN-nah

pesto alla genovese: basil, pine nuts, and garlic
PEHS-toh AHL-lah jeh-noh-VEH-zeh

pomodoro: tomato
poh-moh-DOH-roh

puttanesca: tomato, anchovy, garlic, and capers
poot-tah-NEHS-kah

quattro formaggi: four cheeses
KWAHT-troh fohr-MAHD-jee

ragù: ground meat and tomato
rah-GOO

vongole: clams and tomato
VOHN-goh-leh

RISOTTO (RISOTTI)

risotto ai funghi: with parmesan and porcini mushrooms
ree-ZOHT-toh eye FOON-gy

risotto alla milanese: with saffron and parmesan cheese, Milanese style
ree-ZOHT-toh AHL-lah mee-lah-NEH-zeh

risotto ai frutti di mare: with seafood
ree-ZOHT-toh eye FROOT-tee dee MAH-reh

MEAT (CARNE)

abbacchio alla romana: roasted spring lamb, Roman style
ahb-BAHK-kyoh AHL-lah roh-MAH-nah

bollito misto: boiled cuts of beef and veal served with a sauce
bohl-LEE-toh MEES-toh

carpaccio: thin slices of raw meat—beef, horse, or swordfish—
dressed with olive oil and lemon juice
kahr-PAHT-choh

cinghiale: wild boar, usually braised or roasted
cheen-GYAH-leh

involtini: thin slices of meat filled and rolled
een-vohl-TEE-nee

spezzatino: meat stew
speht-sah-TEE-noh

spiedini: skewered chunks of meat (also available with seafood)
spyeh-DEE-nee

BEEF (MANZO)

bistecca alla fiorentina: grilled T-bone steak, Florentine style
bees-TEHK-kah AHL-lah fyoh-rehn-TEE-nah

bresaola: air-dried beef, served in thin slices
breh-ZAH-oh-lah

osso buco: braised veal shank
OHS-soh BOO-koh

peposo: peppery stew
peh-POH-zoh

stracotto: beef stew with vegetables
strah-KOHT-toh

stufato: stew
stoo-FAH-toh

ORGAN MEATS (INTERIORA)

busecca alla milanese: milanese tripe (beef stomach) soup
boo-SEHK-kah AHL-lah mee-lah-NEH-zeh

cervello al burro nero: brains in black-butter sauce
chehr-VEHL-loh ahl BOOR-roh NEH-roh

fegato alla veneziana: calf's liver fried with onions, Venetian style
FEH-gah-toh AHL-lah veh-neh-TSYAH-nah

VEAL (VITELLO)

bocconcini: stewed or braised chunks
bohk-kohn-CHEE-nee

cima alla genovese: flank steak Genoese style, filled with egg, and
   vegetables
CHEE-mah AHL-lah jeh-noh-VEH-zeh

costoletta alla milanese: breaded cutlet, Milanese style
kohs-toh-LEHT-tah AHL-lah mee-lah-NEH-zeh

lombata di vitello: loin (recipes vary)
lohm-BAH-tah dee vee-TEHL-loh

piccata al Marsala: thin cutlets cooked in Marsala (sweet wine) sauce
peek-KAH-tah ahl mahr-SAH-lah

saltimbocca: veal and ham rolls
sahl-teem-BOHK-kah

scaloppina alla Valdostana: cutlet filled with cheese and ham,
   Valdostana (alpine) style, typically with fontina cheese
skah-lohp-PEE-nah AHL-lah vahl-dohs-TAH-nah

vitello tonnato: cold slices of boiled veal served with tuna-
   mayonnaise sauce
vee-TEHL-loh tohn-NAH-toh

PORK (MAIALE)

arista di maiale: roast loin
ah-REES-tah dee mah-YA-leh

braciola: grilled chop
brah-CHOH-lah

zampone: sausage stuffed pig's trotter
dzahm-POH-neh

POULTRY (POLLAME)

pollo alla cacciatora: chicken braised with mushrooms in tomato sauce, hunter's style
POHL-loh AHL-lah kaht-chah-TOH-rah

pollo alla diavola: chicken chargrilled with lemon and pepper
POHL-loh AHL-lah DYAH-voh-lah

pollo al mattone: chicken grilled with herbs under a brick
POHL-loh ahl maht-TOH-neh

FISH AND SEAFOOD (PESCE E FRUTTI DI MARE)

For more fish and seafood, see p102.

anguilla alla veneziana: eel cooked in tomato sauce, Venetian style
ahn-GWEEL-lah AHL-lah veh-neh-TSYAH-nah

aragosta: lobster
ah-rah-GOHS-tah

baccalà: stockfish or salt cod
bahk-kah-LAH

cacciucco alla livornese: tomato seafood chowder, Livorno style
kaht-CHOOK-koh AHL-lah lee-vohr-NEH-zeh

cozze ripiene: stuffed mussels
KOHT-seh ree-PYEH-neh

gamberi grigliati: grilled shrimp with garlic
GAHM-beh-ree greel-LYAH-tee

fritto misto: mixed fried fish
FREET-toh MEES-toh

nero di seppie e polenta: baby squid in ink squid sauce served with polenta (corn meal)
NEH-roh dee SEHP-pyeh eh poh-LEHN-tah

pesci al cartoccio: fish baked in parchment paper
PEH-shee ahl kahr-TOHT-choh

SIDE DISHES (CONTORNI)

piselli al prosciutto: peas with ham
pee-ZEHL-lee ahl proh-SHOOT-toh

frittata: omelette
freet-TAH-tah

SWEET BREADS / DESSERTS / SWEETS (DOLCI)

amaretti: almond macaroons
ah-mah-REHT-tee

biscotti: cookies
bees-KOHT-tee

cannoli: crispy pastry rolls filled with sweetened ricotta and candied
  fruit
kahn-NOH-lee

cassata alla siciliana: traditional Sicilian cake with ricotta,
  chocolate, and candied fruit
kahs-SAH-tah AHL-lah see-chee-LYAH-nah

gelato: Italian ice cream
jeh-LAH-toh

granita: frozen fruit-juice slush
grah-NEE-tah

panettone: Christmas sweet bread with raisins and candied
  citrus peel
pah-neht-TOH-neh

panforte: almond, candied citrus, spices, and honey cake
pahn-FOHR-teh

panna cotta: molded chilled cream pudding
PAHN-nah KOHT-ta

semifreddo: soft and airy, partially frozen ice cream
seh-mee-FREHD-doh

tartufo: chocolate ice cream dessert
tahr-TOO-foh

tiramisù: mascarpone, ladyfingers, and coffee dessert
tee-rah-mee-SOO

torta: cake
TOHR-tah

  della nonna: custard shortcrust with pine nuts
  DEHL-lah NOHN-na

  alle mele: with apples
  AHL-leh MEH-le

  al limone: with lemon curd
  ahl lee-MOH-neh

alle fragole: with fresh strawberries
AHL-leh FRAH-goh-leh

ai frutti di bosco: with mixed berries
eye FROOT-tee dee BOHS-koh

zabaione: warm custard with Marsala wine
dzah-bah-YOH-neh

zuccotto: spongecake filled with fresh cream, chocolate, candied
fruit, and liqueur
dzook-KOHT-toh

zuppa inglese: a type of English trifle (spongecake layered with
cream and fruit)
DZOOP-pah een-GLEH-zeh

formaggi: cheeses
fohr-MAHD-jee

asiago: nutty-flavored hard cheese
ah-ZYAH-goh

mozzarella di bufala: buffalo mozzarella
moht-tsah-REHL-lah dee BOO-fah-lah

formaggio al latte di pecora: ewe's milk cheese
fohr-MAHD-joh ahl LAHT-teh dee PEH-koh-rah

fontina: semi-soft, melting cheese
fohn-TEE-nah

formaggio al latte di capra: goat's milk cheese
fohr-MAHD-joh ahl LAHT-teh dee KAH-prah

gorgonzola: pungent blue cheese
gohr-gohn-DZOH-lah

mascarpone: rich, dessert cream cheese
mahs-kahr-POH-neh

mozzarella: fresh, soft cow's milk cheese
moht-tsah-REHL-lah

parmigiano: Parmesan
pahr-mee-JAH-noh

provolone: cow's milk cheese, sweet to spicy
proh-voh-LOH-neh

ricotta: fresh, soft and mild cheese
ree-KOHT-tah

taleggio: mild dessert cheese
tah-LEHD-joh
frutta: fruit
FROOT-tah
For a full listing of fruit, see p103.

BUYING GROCERIES

Many Italians shop at **mercati** (open-air markets) or neighborhood specialty stores. Because fresh ingredients are so essential to their cuisine, they often purchase just a day or two's worth of groceries.

AT THE SUPERMARKET

Which aisle has _____	**In quale corsia è / sono _____** *een KWAH-leh kohr-SEE-ah EH / SOH-noh*
spices?	**le spezie?** *leh SPEH-tsyeh*
toiletries?	**gli articoli per l'igiene personale?** *lyee ahr-TEE-koh-lee pehr lee-JEH-neh pehr-soh-NAH-leh*
paper plates and napkins?	**i piatti e tovaglioli di carta?** *ee PYAHT-tee eh toh-vahl-LYOH-lee dee KAHR-tah*
canned goods?	**i cibi in scatola?** *ee CHEE-bee een SKAH-toh-lah*
snack food?	**gli spuntini?** *lyee spoon-TEE-nee*
baby food?	**il cibo per neonati?** *eel CHEE-boh pehr neh-oh-NAH-tee*
water?	**l'acqua?** *LAHK-wah*
juice?	**i succhi?** *ee SOOK-kee*
bread?	**il pane?** *eel PAH-neh*

cheese?	**i formaggi?**
	ee fohr-MAHD-jee
fruit?	**la frutta?**
	lah FROOT-tah
cookies?	**i biscotti?**
	ee bees-KOHT-tee

AT THE BUTCHER SHOP

Do you sell _____	**Ha _____**
	AH
fresh beef?	**del manzo?**
	dehl MAHN-tsoh
fresh pork?	**del maiale?**
	dehl mah-YA-leh
fresh lamb?	**dell'agnello?**
	dehl-ahn-NYEHL-loh
I would like a cut of _____	**Vorrei un taglio di _____**
	vohr-RAY oon TAHL-lyoh dee
tenderloin.	**filetto.**
	fee-LEHT-toh
T-bone.	**fiorentina.**
	fyoh-rehn-TEE-nah
brisket.	**punta di petto.**
	POON-tah dee PEHT-toh
rump roast.	**scamone.**
	skah-MOH-neh
rump.	**girello.**
	jee-REHL-loh
chops.	**braciole.**
	brah-CHOH-leh
filet.	**filetto.**
	fee-LEHT-toh
Thick / Thin cuts, please.	**Tagli spessi / sottili, per favore.**
	TAHL-lyee SPEHS-see / SOT-tee-lee
	pehr fah-VOH-reh

Please trim the fat.	**Tolga il grasso, per favore.** *TOHL-gah eel GRAHS-soh pehr* *fah-VOH-reh*
Do you have any sausage?	**Ha della salsiccia?** *AH DEHL-lah sahl-SEET-chah*

AT THE FISHMONGER

Is the _____ fresh?	**È fresco _____?** *EH FREHS-koh*
fish	**È fresco il pesce?** *EH FREHS-koh eel PEH-sheh*
flounder	**È fresca la passera?** *EH FREHS-kah lah PAHS-seh-rah*
sea bass	**È fresco il branzino?** *EH FREHS-koh eel BRAHN-* *dzee-noh*
shark	**È fresco il palombo?** *EH FREHS-koh eel pah-* *LOHM-boh*
seafood	**Sono freschi i frutti di mare?** *SOH-noh FREHS-kee ee* *FROOT-tee dee MAH-reh*
clams	**Sono fresche le vongole?** *SOH-noh FREHS-kee leh VOHN-* *goh-leh*
octopus	**È fresco il polpo?** *EH FREHS-koh eel POHL-poh*
oysters	**Sono fresche le ostriche?** *SOH-noh FREHS-kee leh OHS-* *tree-keh*
shrimp	**Sono freschi i gamberi?** *SOH-noh FREHS-kee ee GAHM-* *beh-ree*
squid	**Sono freschi i calamari?** *SOH-noh FREHS-kee ee kah-lah-* *MAH-ree*

May I smell it?	**Posso sentire l'odore?**
	POHS-soh sehn-TEE-reh loh-DOH-reh
Would you please ____	**Me lo può ____ per favore?**
	meh loh PWOH ____ pehr fah-VOH-reh
clean it?	**pulire?**
	poo-LEE-reh
filet it?	**sfilettare?**
	sfee-leht-TAH-reh
debone it?	**Mi può togliere le lische?**
	Mee PWOH TOHL-lyeh-reh leh LEES-keh
remove the head and tail?	**Mi può togliere testa e coda?**
	Mee PWOH TOHL-lyeh-reh TEHS-tah eh KOH-dah

AT THE PRODUCE STAND / MARKET

Fruits

apple	**mela**
	MEH-lah
banana	**banana**
	bah-NAH-nah
grapes (green, red)	**uva (bianca, nera)**
	OO-vah BYAHN-kah NEH-rah
orange	**arancia (pl. arance)**
	ah-RAHN-chah
lemon	**limone**
	lee-MOH-neh
lime	**limetta**
	lee-MEHT-tah
cantaloupe, melon	**melone**
	meh-LOH-neh
mango	**mango**
	MAHN-goh
watermelon	**anguria**
	ahn-GOO-ryah

DINING

honeydew	**melone verde** *meh-LOH-neh VERH-deh*
cherry	**ciliegia** *chee-LYEH-jah*
peach	**pesca** *PEHS-kah*
apricot	**albicocca** *ahl-bee-KOHK-kah*
strawberry	**fragola** *FRAH-goh-lah*
wild strawberry	**fragolina di bosco** *frah-goh-LEE-nah dee BOHS-koh*
blueberry (European)	**mirtillo** *meer-TEEL-loh*
kiwi	**kiwi** *kiwi*
pineapple	**ananas** *AH-nah-nahs*
blackberry	**mora** *MOH-rah*
citron	**cedro** *CHEH-droh*
coconut	**cocco** *KOHK-koh*
fig	**fico** *FEE-koh*
grapefruit	**pompelmo** *pohm-PEHL-moh*
guava	**guava** *GWAH-vah*
gooseberry	**uvaspina** *oo-vah-SPEE-nah*
blood orange	**arancia sanguinella** *ah-RAHN-chah sahn-gwee-NEHL-lah*
papaya	**papaia** *pah-PAH-yah*

pear	**pera**
	PEH-rah
plum	**prugna**
	PROON-nyah
yellow plum	**prugna gialla**
	PROON-nyah JAHL-lah
prune	**prugna secca**
	PROON-nyah SEHK-kah
raspberry	**lampone**
	lahm-POH-neh
tangerine	**mandarino**
	mahn-dah-REE-noh

Vegetables

artichoke	**carciofo**
	kahr-CHOH-foh
green asparagus	**asparago verde**
	ahs-PAH-rah-goh VERH-deh
white asparagus	**asparago bianco**
	ahs-PAH-rah-goh BYAHN-koh
avocado	**avocado**
	ah-voh-KAH-doh
beans	**fagioli**
	fah-JOH-lee
green beans	**fagiolini**
	fah-joh-LEE-nee
bamboo shoots	**germogli di bamboo**
	jehr-MOHL-lyee dee bahm-BOO
bean sprout	**germoglio di soia**
	jehr-MOHL-lyoh dee SOH-yah
broccoli	**broccoli**
	BROHK-koh-lee
cabbage	**cavolo**
	KAH-voh-loh

DINING

carrot	**carota**
	kah-ROH-tah
cauliflower	**cavolfiore**
	kah-vohl-FYOH-reh
celery	**sedano**
	SEH-dah-noh
corn	**mais**
	mice
cucumber	**cetriolo**
	cheh-tree-OH-loh
garlic	**aglio**
	AHL-lyoh
eggplant / aubergine	**melanzana**
	meh-lahn-TSAH-nah
fennel	**finocchio**
	fee-NOHK-kyoh
lettuce	**lattuga**
	laht-TOO-gah
arugula	**rucola**
	ROO-koh-lah
radicchio	**radicchio**
	rah-DEEK-kyoh
mushrooms	**funghi**
	FOON-gy
white truffles	**tartufi bianchi**
	tahr-TOO-fee BYAHN-kee
black truffles	**tartufi neri**
	tahr-TOO-fee NEH-ree
nettles	**ortiche**
	ohr-TEE-keh

onion	**cipolla**
	chee-POHL-lah
green olives	**olive verdi**
	oh-LEE-veh VERH-dee
black olives	**olive nere**
	oh-LEE-veh NEH-reh
peas	**piselli**
	pee-ZEHL-lee
peppers	**peperoni**
	peh-peh-ROH-nee
red	**rossi**
	ROHS-see
yellow	**gialli**
	JAHL-lee
green	**verdi**
	VERH-dee
hot	**peperoncino**
	peh-peh-rohn-CHEE-noh
potato	**patata**
	pah-TAH-tah
pumpkin, squash	**zucca**
	DZOOK-kah
sorrel	**acetosella**
	ah-cheh-toh-ZEHL-lah
spinach	**spinaci**
	spee-NAH-chee
tomato	**pomodoro**
	poh-moh-DOH-roh
yam	**patata dolce**
	pah-TAH-tah DOHL-cheh
zucchini	**zucchina**
	dzook-KEE-nah

DINING

Fresh herbs and spices

anise	**anice** *AH-nee-cheh*
basil	**basilico** *bah-ZEE-lee-koh*
bay leaf	**alloro** *ahl-LOH-roh*
black pepper	**pepe nero** *PEH-peh NEH-roh*
clove	**chiodi di garofano** *KYOH-dee dee gah-ROH-fah-noh*
dill	**aneto** *ah-NEH-toh*
garlic	**aglio** *AHL-lyoh*
marjoram	**maggiorana** *mahd-joh-RAH-nah*
oregano	**origano** *oh-REE-gah-noh*
paprika	**paprica** *PAH-pree-kah*
parsley	**prezzemolo** *preht-TSEH-moh-loh*
rosemary	**rosmarino** *rohz-mah-REE-noh*
saffron	**zafferano** *dzahf-feh-RAH-noh*
sage	**salvia** *SAHL-vyah*
salt	**sale** *SAH-leh*

sugar	**zucchero**
	DZOOK-keh-roh
thyme	**timo**
	TEE-moh

AT THE DELI

What kind of salad is that?
Che tipo di insalata è quella?
keh TEE-poh dee een-sah-LAH-tah EH KWEHL-lah

What type of cheese is that?
Che tipo di formaggio è quello?
keh TEE-poh dee fohr-MAHD-joh EH KWEHL-loh

What type of bread is that?
Che tipo di pane è quello?
keh TEE-poh dee PAH-neh EH KWEHL-loh

May I have some of that please?
Posso avere un po' di quello, per favore?
POHS-soh ah-VEH-reh oon POH dee KWEHL-loh pehr fah-VOH-reh

Is the salad fresh?
Questa insalata è fresca?
KWEHS-tah een-sah-LAH-tah EH FREHS-kah

I'd like _____, please.
Vorrei _____, per favore.
vohr-RAY _____ pehr fah-VOH-reh

a sandwich
un panino
oon pah-NEE-noh

a salad
un'insalata
oon-een-sah-LAH-tah

tuna salad
un'insalata di tonno
oon-een-sah-LAH-tah dee TOHN-noh

chicken salad
un'insalata di pollo
oon-een-sah-LAH-tah dee POHL-loh

roast beef	**del rosbif**
	dehl ROHZ-beef
shrimp cocktail	**un cocktail di gamberetti**
	oohn KOHK-tayl dee gahm-beh-REHT-tee
ham	**del prosciutto**
	dehl proh-SHOOT-toh
mustard	**della senape**
	DEHL-lah SEH-nah-peh
mayonnaise	**della maionese**
	DEHL-lah mah-yoh-NEH-zeh
a package of tofu	**un pacchetto di tofu**
	oon pahk-KEHT-toh dee TOH-foo
a pickle	**dei cetrioli sott'aceto**
	deh-ee cheh-tree-OH-lee sot-tah-CHEH-toh
I'd like that cheese.	**Mi piace quel formaggio.**
	mee PYAH-cheh kwehl fohr-MAHD-joh
Is that smoked?	**È affumicato -a?**
	EH ahf-FOO-mee-kah-toh -tah
gram	**grammo -i**
	GRAHM-moh -mee
kilo	**chilo -i**
	KEE-loh -lee
half-kilo	**mezzo chilo**
	MEH-dzo KEE-loh
quarter-kilo	**250 grammi**
	DOO-eh CHEHN-toh cheen-KWAN-tah GRAHM-mee

CHAPTER FIVE

SOCIALIZING

Italians are a gregarious and curious people. They often initiate conversations with foreigners (some of which can seem quite personal). In this family-oriented country, a solo traveler is often viewed with sympathy; don't assume that every invitation is a scam or flirtation. Here you'll find the language to make new friends.

GREETINGS

Hello.	**Salve.**
	SAHL-veh
Good morning.	**Buon giorno.**
	bwon JOHR-noh
Good afternoon.	**Buon pomeriggio.**
	bwon poh-meh-REED-joh
Good evening.	**Buona sera.**
	BWOH-nah SEH-rah
Good night.	**Buona notte.**
	BWOH-nah NOHT-teh
How are you?	**Come va?**
	KOH-meh vah
Fine, thanks.	**Bene, grazie.**
	BEH-neh GRAH-tsyeh
And you?	**E tu / lei / voi?**
	eh too / lay / voy
I'm exhausted.	**Sono esausto -a.**
	SOH-noh eh-ZOWS-toh -tah
I have a headache.	**Ho il mal di testa.**
	OH eel mahl dee TEHS-tah
I'm terrible.	**Sto male.**
	stoh MAH-leh
I have a cold.	**Ho il raffreddore.**
	OH eel rahf-frehd-DOH-reh

Listen Up: Common Greetings

Ciao. *CHAH-oh*	Hi / Bye.
Salve. *SAHL-veh*	Hello.
È un piacere. *EH oon pyah-CHEH-reh*	It's a pleasure.
Piacere. *pyah-CHEH-reh*	How do you do / nice to meet you.
Molto piacere. *MOHL-toh pyah-CHEH-reh*	Delighted.
Come va? *KOH-meh vah*	How's it going?
Arrivederci. *ahr-ree-veh-DEHR-chee*	Goodbye.
A più tardi. *ah PYOO TAHR-dee*	See you later.
Ci vediamo. *chee veh-DYAH-moh*	See you later.

THE LANGUAGE BARRIER

I don't understand.	**Non capisco.** *nohn kah-PEES-koh*
Please speak more slowly.	**Parli più lentamente, per favore.** *PAHR-lee PYOO lehn-tah-MEHN-teh pehr fah-VOH-reh*
Please speak louder.	**Parli a voce più alta, per favore.** *PAHR-lee ah VOH-cheh PYOO AHL-tah pehr fah-VOH-reh*
Do you speak English?	**Parla inglese?** *PAHR-lah een-GLEH-seh*
I speak ____ better than Italian.	**Parlo ____ meglio dell'italiano.** *PAHR-loh ____ MEHL-lyoh DEHL-lee-tah-LYAH-noh*

Please spell that.	**Mi può dire come si scrive, per favore?** *Mee PWOH DEE-reh KOH-meh see SKREE-veh, pehr fah-VOH-reh*
Please repeat that?	**Me lo ripete, per favore?** *meh loh ree-PEH-teh pehr fah-VOH-reh*
How do you say _____?	**Come si dice _____?** *KOH-meh see DEE-cheh*
Would you show me that in this dictionary?	**Me lo mostra sul dizionario?** *Meh loh MOHS-trah suhl dee-tsyoh-NAH-ryoh*

Curse Words

Here are some common curse words.

merda *MEHR-dah*	shit
figlio di puttana *FEEL-lyoh dee poot-TAH-nah*	son of a bitch (literally "son of a whore")
stronzo *STROHN-tsoh*	jerk (literally "turd", quite common)
Cazzo! *KAHT-soh*	Damn! (literally "dick," stronger than "damn," very frequently used)
Che cazzo vuoi? *keh KAHT-soh VWOH-ee*	What the hell do you want?
culo *KOO-loh*	ass (meaning the behind)
incasinato *een-kah-zee-NAH-toh*	screwed up
Vaffanculo! *vahf-fahn-KOO-loh*	Fuck off!

GETTING PERSONAL

Italians are generally friendly, yet more formal than Americans. Remember to use the formal lei (third-person singular) until given permission to employ the more familiar tu.

INTRODUCTIONS

What is your name?	**Come si chiama?** *KOH-meh see KYAH-mah*
My name is ____.	**Mi chiamo ____.** *mee KYAH-moh*
I'm pleased to meet you.	**Piacere di conoscerla.** *pyah-CHEH-reh dee koh-NOSH-ehr-lah*
May I introduce my ____	**Posso presentarle mio -a ____** *POHS-soh preh-zehn-TAHR-leh MEE-oh -ah*
wife?	**moglie?** *MOHL-lyeh*
husband?	**marito?** *mah-REE-toh*
son / daughter?	**figlio -a** *FEEL-lyoh -lyah*
friend?	**il / la mio -a amico-a?** *eel / lah MEE-oh -ah ah-MEE-koh -kah*
boyfriend / girlfriend?	**il / la mio -a ragazzo -a?** *eel / lah MEE-oh -ah rah-GAHT-soh -sah*
How is your ____	**Come sta -anno il / la suo -a ____** *KOH-meh stah -STAHN-noh eel / lah SOO-oh -ah*
family?	**famiglia?** *fah-MEEL-lyah*
mother?	**madre?** *MAH-dreh*
father?	**padre?** *PAH-dreh*

SOCIALIZING

brother / sister?	**suo fratello / sua sorella?** *SOO-oh frah-TEHL-loh / SOO-ah soh REHL-lah*
neighbor?	**vicino -a?** *vee-CHEE-noh -nah*
boss?	**capo?** *KAH-poh*
cousin?	**suo cugino / sua cugina?** *SOO-oh koo-JEE-noh -nah / SOO-ah koo-JEE-noh -nah*
aunt / uncle?	**suo zio / sua zia?** *SOO-oh DZEE-oh / SOO-ah DZEE-ah*
fiancée / fiancé?	**fidanzato -a?** *fee-dahn-TSAH-toh -tah*
partner?	**partner?** *partner*
niece / nephew?	**suo nipote / sua nipote?** *SOO-oh nee-POH-teh / SOO-ah nee-POH-teh*
How are your ____	**Come sta -anno i suoi ____** *KOH-meh stah -STAHN-noh ee soo-OH-ee*
children?	**bambini / figli (if older than teens)** *bahm-BEE-nee / FEEL-lyee*
parents?	**genitori?** *jeh-nee-TOH-ree*
grandparents?	**nonni?** *NOHN-nee*
grandchildren?	**nipoti?** *nee-POH-tee*

Dos and Don'ts.

Don't refer to your parents as **i parenti** (*ee pah-REHN-tee*), which means relatives. Do call them **i genitori** (*ee jeh-nee-TOH-ree*).

Are you married?	**È sposato -a?** *EH spoh-ZAH-toh -tah*
I'm married.	**Sono sposato -a.** *SOH-noh spoh-ZAH-toh -tah*
I'm single.	**Non sono sposato -a.** *nohn SOH-noh spoh-ZAH-toh -tah*
I'm divorced.	**Sono divorziato -a.** *SOH-noh dee-vohr-TZYAH-toh -tah*
I'm a widow / widower.	**Sono vedovo -a.** *SOH-noh VEH-doh-voh -vah*
We're separated.	**Siamo separati.** *SYAH-moh seh-pah-RAH-tee*
I live with my boyfriend / girlfriend.	**Vivo con il mio / la mia ragazzo -a.** *VEE-voh kohn eel MEE-oh / lah MEE-ah rah-GAHT-soh -sah*
I live with a roommate.	**Vivo con un coinquilino.** *VEE-voh kohn oon koh-een-kwee-LEE-noh*
How old are you?	**Quanti anni ha?** *KWAHN-tee AHN-nee AH*
How old are your children?	**Quanti anni hanno i suoi bambini?** *KWAHN-tee AHN-nee AHN-noh ee soo-OH-ee bahm-BEE-nee*
Wow, that's very young.	**Ah, è molto giovane.** *AH EH MOHL-toh JOH-vah-neh*
No, you're not! You're much younger.	**No, davvero! Lei è molto più giovane.** *noh dahv-VEH-roh lay EH MOHL-toh PYOO JOH-vah-neh*
Your wife / daughter is beautiful.	**Sua moglie / figlia è bellissima.** *SOO-ah MOHL-lyeh / FEEL-lyah EH behl-LEES-see-mah*
Your husband / son is handsome.	**Suo marito / figlio è molto bello.** *SOO-oh mah-REE-toh / FEEL-lyoh EH MOHL-toh BEHL-loh*
What a beautiful baby!	**Che bel -la bambino -a!** *keh behl -lah bahm-BEE-noh -nah*

Are you here on business?	**È qui per lavoro?** *EH kwee pehr lah-VOH-roh*
I am vacationing.	**Sono in vacanza.** *SOH-noh een vah-KAHN-tsah*
I'm attending a conference.	**Sono qui per una conferenza.** *SOH-noh KWEE pehr OO-nah kohn-feh-REHN-tsah*
I'm traveling with my husband/wife.	**Sono qui con mio marito / mia moglie** *SOH-noh KWEE kohn mee-oh mah-REE-toh/MEE-ah MOHL-lyeh*
How long are you staying?	**Quanto tempo si ferma?** *KWAHN-toh TEHM-poh see FEHR-mah*
I'm a student.	**Sono uno studente / una studentessa.** *SOH-noh OO-noh stoo-DEHN-teh / OO-nah stoo-DEHN-tehs-sah*
What are you studying?	**Cosa studia?** *KOH-zah STOO-dyah*
Where are you from?	**Da dove viene?** *Dah DOH-veh VYEH-neh*

PERSONAL DESCRIPTIONS

afro	**capigliatura africana** *kah-peel-lyah-TOO-rah ah-free-KAH-nah*
blonde	**biondo -a** *BYOHN-doh -dah*
brunette	**castano -a** *KAHS-tah-noh -nah*
redhead	**rosso -a** *ROHS-soh -sah*
curly hair	**capelli ricci** *kah-PEHL-lee REET-chee*
kinky hair	**capelli crespi** *kah-PEHL-lee KREHS-pee*

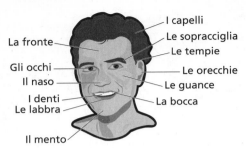

I capelli
Le sopracciglia
Le tempie
La fronte
Gli occhi
Il naso
Le orecchie
Le guance
I denti
Le labbra
La bocca
Il mento

long hair	**capelli lunghi** *kah-PEHL-lee LOON-gy*
short hair	**capelli corti** *kah-PEHL-lee KOHR-tee*
straight hair	**capelli lisci** *kah-PEHL-lee LEE-shee*
black	**nero -a** *NEH-roh -rah*
pale	**pallido -a** *PAHL-lee-doh -dah*
mocha-skinned	**dalla pelle color caffè** *DAHL-lah PEHL-leh koh-LOHR* *kahf-FEH*
olive-skinned	**con la pelle olivastra** *kohn lah PEHL-leh oh-lee-* *VAHS-trah*
tanned	**abbronzato -a** *ahb-brohn-DZAH-toh -tah*
white	**bianco -a** *BYAHN-koh -kah*
Asian	**asiatico -a** *ah-ZYAH-tee-koh -kah*

biracial	**meticcio -a**
	meh-TEET-choh -chah
African-American	**afroamericano -a**
	AH-froh-ah-meh-ree-KAH-noh -nah
caucasian	**caucasico -a**
	kow-KAH-zee-koh -kah
tall	**alto -a**
	AHL-toh -tah
short	**basso -a**
	BAHS-soh -sah
thin	**magro -a**
	MAH-groh -grah
fat	**grasso -a**
	GRAHS-soh -sah
blue eyes	**occhi azzurri**
	OHK-kee ahd-DZOOR-ree
brown eyes	**occhi castani**
	OHK-kee kahs-TAH-nee
green eyes	**occhi verdi**
	OHK-kee VERH-dee
hazel eyes	**occhi nocciola**
	OHK-kee noht-CHOH-lah
eyebrows	**sopracciglia**
	soh-praht-CHEEL-lyah
eyelashes	**ciglia**
	CHEEL-lyah
freckles	**lentiggini**
	lehn-TEED-jee-nee
moles	**nei**
	nay
face	**viso**
	VEE-zoh

Listen Up: Nationalities

Sono _____	I'm _____
SOH-noh	
brasiliano -a.	Brazilian.
brah-zee-LYAH-noh -nah	
francese.	French.
frahn-CHEH-zeh	
greco -a.	Greek.
GREH-koh -kah	
portoghese.	Portuguese.
pohr-toh-GHEH-zeh	
rumeno -a.	Romanian.
roo-MEH-noh -nah	
russo -a.	Russian.
ROOS-soh -sah	
spagnolo -a.	Spanish.
spahn-NYOH-loh -lah	
svizzero -a.	Swiss.
ZVEET-seh-roh -rah	
tedesco -a.	German.
teh-DEHS-koh -kah	
ungherese.	Hungarian.
oon-gheh-REH-zeh	

For a full list of nationalities, see English / Italian dictionary.

DISPOSITIONS AND MOODS

sad	**triste**
	TREES-teh
happy	**felice**
	feh-LEE-cheh
angry	**arrabbiato -a**
	ahr-rahb-BYAH-toh -tah
tired	**stanco -a**
	STAHN-koh -kah
depressed	**depresso -a**
	deh-PREHS-soh -sah

stressed	**stressato -a** *strehs-SAH-toh -tah*
anxious	**ansioso -a** *ahn-SYOH-zoh -zah*
confused	**confuso -a** *kohn-FOO-zoh -zah*
enthusiastic	**entusiasta** *ehn-too-ZYAHS-tah*

PROFESSIONS

What do you do for a living?	**Che lavoro fa?** *keh lah-VOH-roh fah*
Here is my business card.	**Ecco il mio biglietto da visita.** *EHK-koh eel MEE-oh beel-LYEHT-toh dah VEE-zee-tah*
I am _____	**Faccio _____** *FAH-choh*
an accountant.	**il/la ragioniere -a.** *eel/lah rah-jyoh-NYEH-reh -rah*
an artist.	**l'artista.** *lahr-TEES-tah*
a craftsperson.	**l'artigiano -a.** *lahr-tee-JAH-noh -nah*
a designer.	**lo/la stilista.** *loh/lah stee-LEES-tah*
a doctor.	**il medico.** *eel MEH-dee-koh*
an editor.	**il redattore / la redattrice.** *eel reh-daht-TOH-reh / lah reh-daht-TREE-che*
an educator.	**l'educatore / l'educatrice.** *leh-doo-kah-TOH-reh / leh-doo-kah-TREE-cheh*
an engineer.	**l'ingegnere.** *leen-jehn-NYEH-reh*

a government employee.
l'impiegato statale.
leem-pyeh-GAH-toh stah-TAH-leh

a homemaker.
la casalinga.
lah kah-zah-LEEN-gah

a lawyer.
l'avvocato / l'avvocatessa.
lahv-voh-KAH-toh / lahv-voh-KAH-tess-sah

a military professional.
il militare.
eel mee-lee-TAH-reh

a musician.
il musicista.
eel moo-zee-CHEES-tah

a nurse.
l'infermiere -a.
leen-fehr-MYEH-reh -rah

a salesperson.
il/la commesso -a.
eel/lah kohm-MEHS-soh -sah

a writer.
lo scrittore / la scrittrice.
loh skreet-TOH-reh / lah skreet-TREE-cheh

a web designer.
il web designer.
eel web designer

a computer programmer.
il programmatore informatico.
eel proh-grahm-mah-TOH-reh een-fohr-MAH-tee-koh

DOING BUSINESS

I'd like an appointment.
Vorrei un appuntamento.
vohr-RAY oon ahp-poon-tah-MEHN-toh

I'm here to see ____.
Sono qui per vedere ____.
SOH-noh kwee pehr veh-DEH-reh

May I photocopy this?
Posso fotocopiare questo?
POHS-soh foh-toh-koh-PYAH-reh KWEHS-toh

May I use my laptop?
Posso usare il mio portatile?
POHS-soh oo-ZAH-reh eel mee-oh pohr-TAH-tee-leh

What's the password?	**Qual è la password?** *kwah-LEH lah password*
May I access the Internet?	**Posso accedere a Internet?** *POHS-soh aht-CHEH-deh-reh ah internet*
May I send a fax?	**Posso inviare un fax?** *POHS-soh een-VYAH-reh oon fax*
May I use the phone?	**Posso usare il telefono?** *POHS-soh oo-ZAH-reh eel teh-LEH-foh-noh*
Do you have a Wi-Fi network?	**Ha una rete Wi-Fi?** *Ah oo-nah REH-teh Wi-Fi*

PARTING WAYS

Keep in touch.	**Teniamoci in contatto.** *teh-NYAH-moh-chee een kohn-TAHT-toh*
Please write or email.	**Scriva o invii un'email.** *SKREE-vah oh een-VEE-ee oon e-mail*
Here's my phone number. Call me.	**Questo è il mio numero di telefono. Mi chiami.** *KWEH-stoh EH eel MEE-oh NOO-meh-roh dee teh-LEH-poh-noh. mee KYAH-mee.*
May I have your phone number / email, please?	**Posso avere il suo numero di telefono / indirizzo e-mail, per favore?** *POHS-soh ah-VEH-reh eel SOO-oh NOO-meh-roh dee teh-LEH-foh-noh / een-dee-REET-soh e-mail pehr fah-VOH-reh*
May I have your card?	**Posso avere il suo biglietto da visita?** *POHS-soh ah-VEH-reh eel SOO-oh beel-LYEHT-toh dah VEE-zee-tah*
Are you on Facebook / Twitter?	**E' su Facebook / Twitter?** *EH soo FAYS-book/TWEET-tehr*

TOPICS OF CONVERSATION

As in the United States, Europe, or anywhere in the world, the weather and current affairs are common conversation topics.

THE WEATHER

It's _____
È _____
EH

Is it always so _____
È sempre così _____
EH SEHM-preh koh-ZEE

cloudy?
nuvoloso?
noo-voh-LOH-zoh

humid?
umido?
OO-mee-doh

warm?
caldo?
KAHL-doh

cool?
fresco?
FREHS-koh

rainy?
piovoso?
pyoh-VOH-zoh

windy?
C'è sempre così tanto vento?
ch-EH SEHM-preh koh-ZEE
TAHN-toh VEHN-toh

sunny?
sole?
SOH-leh

Do you know the weather forecast for tomorrow?
Quali sono le previsioni del tempo per domani?
KWAH-lee SOH-noh leh preh-vee-ZYOH-nee dehl TEHM-poh pehr doh-MAH-nee

THE ISSUES

What do you think about _____

Che ne pensa _____
keh neh PEHN-sah

the government?

del governo?
dehl goh-VEHR-noh

democracy?

della democrazia?
DEHL-lah deh-moh-krah-TSEE-ah

socialism?

del socialismo?
dehl soh-chah-LEEZ-moh

the environment?

dell'ambiente?
DEHL-lahm-BYEHN-teh

women's rights?

dei diritti delle donne?
day dee-REET-tee DEHL-leh DOHN-neh

gay rights?

dei diritti degli omosessuali?
day dee-REET-tee DEHL-lyee oh-moh-sehs-SWAH-lee

the economy?

dell'economia?
dehl-leh-koh-noh-MEE-ah

What political party do you belong to?

A quale partito politico appartiene?
ah KWAH-leh pahr-TEE-toh poh-LEE-tee-koh ahp-pahr-TYEH-neh

What did you think of the election?

Come le sono sembrate le elezioni?
KOH-meh leh SOH-noh sehm-BRAH-teh leh eh-leh-TSYOH-nee

What do you think of the war in _____?

Cosa pensa della guerra in _____?
KOH-zah PEHN-sah DEHL-lah GWEHR-rah een

RELIGION

Do you go to church / temple / mosque?	**Lei frequenta una chiesa / un tempio / una moschea?** *lay freh-KWEHN-tah OO-nah KYEH-zah / oon TEHM-pyoh / OO-nah mohs-KEH-ah*
Are you religious?	**Lei è praticante?** *lay EH prah-tee-KAHN-teh*
I'm _____ / I was raised _____	**Sono _____ / Sono cresciuto -a _____** *SOH-noh _____ / SOH-noh kreh-SHOO-toh -tah*
agnostic.	**agnostico -a.** *ahn-NYOHS-tee-koh -kah*
atheist.	**ateo -a.** *AH-teh-oh -ah*
Buddhist.	**buddista.** *bood-DEES-tah*
Catholic.	**cattolico -a.** *kaht-TOH-lee-koh -kah*
Greek Orthodox.	**greco ortodosso -a.** *GREH-koh ohr-toh-DOHS-soh -sah*
Hindu.	**hindu.** *hindu*
Jewish.	**ebreo -a.** *eh-BREH-oh -ah*
Muslim.	**mussulmano -a.** *moos-sool-MAH-noh -nah*
Protestant.	**protestante.** *proh-tehs-TAHN-teh*
I'm spiritual but I don't attend services.	**Sono credente ma non praticante.** *SOH-noh Kreh-DEHN-teh mah nohn prah-tee-KAHN-teh*

I don't believe in that.	**Non ci credo.**
	nohn chee KREH-doh
That's against my beliefs.	**È contrario alle mie convinzioni.**
	EH kohn-TRAH-ryoh AHL-leh MEE-
	eh kohn-veen-TSYOH-nee
I'd rather not talk about it.	**Preferisco non parlarne.**
	preh-feh-REES-koh nohn pahr-
	LAHR-neh

GETTING TO KNOW SOMEONE

Following are some conversation starters.

MUSICAL TASTES

What kind of music do you like?	**Che tipo di musica le piace?**
	keh TEE-poh dee MOO-zee-kah leh
	PYAH-cheh
I like _____	**Mi piace _____**
	mee PYAH-cheh
rock 'n' roll.	**il rock 'n' roll.**
	eel rock 'n' roll
hip hop.	**l' hip hop.**
	lee hip hop
techno.	**la techno.**
	lah techno
disco.	**la musica da discoteca.**
	lah MOO-zee-kah dah dees-koh
	TEH-kah
classical.	**la musica classica.**
	lah MOO-zee-kah KLAHS-see-kah
jazz.	**il jazz.**
	eel jazz
country and western.	**la musica country e western.**
	lah MOO-zee-kah country eh
	western.

reggae.

il reggae.
eel reggae

calypso.

il calypso.
eel calypso

opera.

l'opera.
LOH-peh-rah

show-tunes / musicals.

le canzoni da musical.
leh kahn-TSOH-nee dah musical

New Age.

la New Age.
lah new age

pop.

il pop.
eel pop

HOBBIES?

What do you like to do in
your spare time?

**Cosa le piace fare nel tempo
libero?**
*KOH-zah leh PYAH-cheh FAH-reh
nehl TEHM-poh LEE-beh-roh*

I like _____

Mi piace _____
mee PYAH-cheh

playing guitar.

suonare la chitarra.
*swoh-NAH-reh lah kee-
TAHR-rah*

piano.

il pianoforte.
eel pyah-noh-FOHR-teh

For other instruments, see the English / Italian dictionary.

painting.

dipingere.
dee-PEEN-jeh-reh

drawing.

disegnare.
dee-zehn-NYAH-reh

dancing.

andare a ballare.
anh-DAH-reh ah bahl-LAH-reh

reading.

leggere.
LEHD-jeh-reh

watching TV.

guardare la TV.
ghwar-DAH-reh lah tee-VOO

blogging.	**scrivere sui blog** *SKREE-veh-reh SOO-ih blog*
shopping.	**fare shopping.** *FAH-reh shopping*
going to the movies.	**andare al cinema.** *anh-DAH-reh ahl CHEE-neh-mah*
hiking.	**fare camminate.** *FAH-reh kahm-mee-NAH-teh*
camping.	**andare in campeggio.** *anh-DAH-reh een kahm-PEHD-joh*
hanging out.	**uscire con gli amici.** *oo-SHEE-reh kohn lyee ah-MEE-chee*
traveling.	**viaggiare.** *vyahd-JAH-reh*
eating out.	**mangiare fuori.** *mahn-JAH-reh FWOH-ree*
cooking.	**cucinare.** *koo-chee-NAH-reh*
sewing.	**cucire.** *koo-CHEE-reh*
sports.	**fare sport.** *FAH-reh sport*
Do you like to dance?	**Vuole ballare?** *VWOH-leh bahl-LAH-reh*
Would you like to go out?	**Vuole uscire?** *VWOH-leh oo-SHEE-reh*
May I buy you dinner sometime?	**Posso portarla fuori a cena qualche volta?** *POHS-soh pohr-TAHR-lah FWOH-ree ah CHEH-nah KWAHL-keh VOHL-tah*
What kind of food do you like?	**Che tipo di cibo le piace?** *keh TEE-poh dee CHEE-boh leh PYAH-cheh*

For a full list of food types, see Dining in Chapter 4.

Would you like to go _____	**Le piacerebbe andare _____** *leh pyah-CHEH-rehb-beh ahn-DAH-reh*
to a movie?	**al cinema?** *ahl CHEE-neh-mah*
to a concert?	**ad un concerto?** *ah-DOON kohn-CHEHR-toh*
to the zoo?	**allo zoo?** *AHL-loh DZOH-oh*
to the beach?	**al mare?** *ahl MAH-reh*
to a museum?	**al museo?** *ahl moo-ZEH-oh*
for a walk in the park?	**a passeggiare nel parco?** *ah pahs-sehd-JAH-reh nehl PAHR-koh*
dancing?	**a ballare?** *ah bahl-LAH-reh*
Would you like to get _____	**Andiamo _____** *ahn-DYAH-moh*
lunch?	**a pranzo?** *ah PRAHN-tsoh*
coffee?	**a prendere un caffè?** *ah PREHN-deh-reh oon kahf-FEH*
dinner?	**a cena?** *ah CHEH-nah*
What kind of books do you like to read?	**Che tipo di libri le piace leggere?** *keh TEE-poh dee LEE-bree leh PYAH-cheh LEHD-jeh-reh*
I like _____	**Mi piacciono _____** *mee PYAHT-choh-noh*
mysteries.	**i gialli.** *ee JAHL-lee*

Westerns.	**i romanzi western.**
	ee roh- MAHN-dzee western
dramas.	**le storie drammatiche.**
	leh STOH-ryeh drahm-MAH-tee-keh
novels.	**i romanzi.**
	ee roh- MAHN-dzee
biographies.	**le biografie.**
	leh byoh-grah-FEE-eh
autobiographies.	**le autobiografie.**
	leh ow-toh-byoh-grah-FEE-eh
romance.	**i romanzi d'amore.**
	ee roh- MAHN-dzee dah-MOH-reh
history.	**i saggi storici.**
	ee SAHD-jee STOH-ree-chee

For dating terms, see Nightlife in Chapter 10.

SOCIALIZING

CHAPTER SIX

MONEY AND COMMUNICATIONS

This chapter covers money, the mail, phone and Internet service, and other tools you need to connect with the outside world.

MONEY

In 2002, Italy entered the Eurozone, changing its currency from lira to euro. The euro is available in seven different bills (5, 10, 20, 50, 100, 200, and 500) and eight separate coins (1, 2, 5, 10, 20, 50 centesimi, 1 and 2 euro denominations). Bills over 20 euros may be difficult to break.

CURRENCIES

I need to exchange money.	**Devo cambiare dei soldi.** *DEH-voh kahm-BYAH-reh DE-hee SOHL-dee*
Do you accept _____	**Accettate _____** *aht-cheht-TAH-teh*
Visa / MasterCard / Discover / American Express / Diners' Club?	Visa / MasterCard / Discover / American Express / Diners' Club?
credit cards?	**carte di credito?** *KAHR-teh dee KREH-dee-toh*
bills?	**banconote?** *bahn-koh-NOH-teh*
coins?	**monete?** *moh-NEH-teh*
checks?	**assegni?** *ahs-SEHN-nyee*
money transfer?	**un bonifico bancario?** *oon boh-NEE-fee-koh bahn-KAH-ree-oh*

May I wire transfer funds here?	**Posso effettuare un bonifico bancario qui?** *POHS-soh ehf-feht-TWAH-reh oon boh-NEE-fee-koh bahn-KAH-ree-oh kwee*
Would you please tell me where to find _____	**Può dirmi dove si trova _____** *Pwoh-DEER-mee DOH-veh see TROH-vah*
a bank?	**una banca?** *OO-nah BAHN-kah*
a credit bureau?	**un ufficio informazioni commerciali?** *oon oof-FEE-choh een-fohr-mah-TSYO-nee kohm-mehr-CHA-lee*
an ATM / cashpoint?	**un bancomat?** *oon BAHN-koh-maht*
a currency exchange?	**un ufficio di cambio?** *oof-FEE-choh dee kahm byoh*
A receipt, please.	**Una ricevuta, per favore.** *OO-nah ree-cheh-VOO-tah pehr fah-VOH-reh*
Would you tell me _____	**Può dirmi qual è _____** *PWOH DEER-mee kwah-LEH*
the exchange rate for dollars to _____?	**il cambio dal dollaro a _____?** *eel KAHM-byoh dahl DOHL-lah-roh ah*
Is there a service charge?	**È necessario pagare una commissione?** *EH neh-chehs-SAH-rio pah-GAH-reh OO-nah kohm-mees-SYOH-neh*

Listen Up: Bank Lingo

Firmi qui, per favore.
*FEER-mee kwee
pehr fah-VOH-reh*

Please sign here.

Ecco la sua ricevuta.
*EHK-koh lah SOO-ah
ree-che-VOO-tah*

Here is your receipt.

Ha un documento d'identità?
*AH oon doh-koo-MEHN-toh
dee-dehn-tee-TAH*

May I see your ID?

Solo contanti.
SOH-loh kohn-TAHN-tee

Cash only.

May I have a cash advance on my credit card?

Posso avere un anticipo sulla mia carta di credito?
POHS-soh ah-VEH-reh oon ahn-TEE-chee-poh SOOL-lah MEE-ah KAHR-tah dee KREH-dee-toh

May I have smaller bills, please.

Banconote più piccole, per favore.
bahn-koh-NOH-teh PYOO PEEK-koh-leh pehr fah-VOH-reh

Can you make change?

Può cambiare?
PWOH kahm-BYAH-reh

I only have bills.

Ho solo banconote.
OH SOH-loh bahn-koh-NOH-teh

Some coins, please.

Degli spiccioli, per favore.
DEHL-lyee SPEET-choh-lee pehr fah-VOH-reh

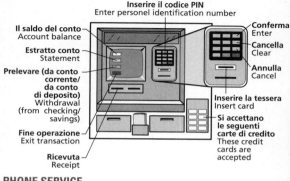

Inserire il codice PIN
Enter personel identification number

Il saldo del conto
Account balance

Estratto conto
Statement

Prelevare (da conto corrente/ da conto di deposito)
Withdrawal (from checking/ savings)

Fine operazione
Exit transaction

Ricevuta
Receipt

Conferma
Enter

Cancella
Clear

Annulla
Cancel

Inserire la tessera
Insert card

Si accettano le seguenti carte di credito
These credit cards are accepted

COMMUNICATIONS

PHONE SERVICE

Where can I buy or rent a mobile / cell phone?	**Dove posso comprare o noleggiare un cellulare?** *DOH-veh POHS-soh kohm-PRAH-reh oh noh-lehd-JAH-reh oon chehl-loo-LAH-reh*
What rate plans do you have?	**Che piani tariffari avete?** *keh PYAH-nee tah-reef-FAH-ree ah-VEH-teh*
May I have a pre-paid phone?	**Posso avere un telefono pre-pagato?** *POHS-soh ah-VEH-reh oon teh-LEH-foh-noh preh-pah-GAH-toh*
Where can I buy a phone card?	**Dove posso comprare una scheda telefonica?** *DOH-veh POHS-soh kohm-PRAH-reh OO-nah skay-dah teh-LEH-foh-nee-kah*
Is data included in the rate?	**Il traffico dati è incluso nella tariffa?** *eel trahf-FEE-koh DAH-tee EH een-KLOO-soh NEHL-lah tah-REF-fah*

I'd like a pay-as-you-go
SIM card.

Vorrei una carta SIM ricaricabile
VOHR-reyh OO-nah KAHR-tah SIM
ree-kah-ree-KAH-bee-leh

May I add more minutes
to my phone card?

Posso aggiungere altri minuti alla
mia scheda telefonica?
POHS-soh ahd-JOON-jeh-reh AHL-
tree mee-NOO-tee AHL-lah MEE-
ah SKEH-dah teh-leh-FOH-nee-kah

MAKING A CALL

May I dial direct?

Posso chiamare direttamente?
POHS-soh kyah-MAH-reh
dee-reht-tah-MEHN-teh

Operator, please.

Il centralino, per favore.
eel cehn-trah-LEE-noh
pehr fah-VOH-reh

I'd like to make
an international call.

Vorrei fare una chiamata
internazionale.
vohr-RAY FAH-reh OO-nah kyah-
MAH-tah een-tehr-nah-tsyoh-
NAH-leh

Fuori servizio

Before you stick your coins or bills in a vending machine,
watch out for the little sign that says **Fuori Servizio** (Out
of Service).

Listen Up: Telephone Lingo

Pronto? / Sì?
PROHN-toh / SEE /

Hello?

Che numero?
keh NOO-meh-roh

What number?

Mi dispiace, la linea è occupata.
mee dee-SPYAH-cheh lah LEE-neh ah EH ohk-koo-PAH-tah

I'm sorry, the line is busy.

Per favore, riagganciare e ricomporre il numero.
pehr fah-VOH-reh ree-ahg-ahg-ghan-CHAH-reh eh ree-kohm-POHR-reh eel NOO-meh-roh

Please, hang up and redial.

Mi dispiace, non risponde nessuno.
mee dee-SPYAH-cheh nohn rees-POHN-deh nehs-SOO-noh

I'm sorry, nobody is answering.

La sua scheda ha ancora dieci minuti a disposizione.
lah SOO-ah SKEH-dah AH ahn-KOH-rah DYEH-chee mee-NOO-tee ah dees-poh-zee-TSYOH-neh

Your card has ten minutes left.

I'd like to make a collect call.	**Vorrei fare una chiamata a carico del ricevente.** *vohr-RAY FAH-reh OO-nah kyah-MAH-tah ah KAH-ree-koh dehl ree-cheh-VEHN-teh*
I'd like to use a calling card.	**Vorrei usare la scheda telefonica.** *vohr-RAY oo-ZAH-reh lah SKEH-dah teh-leh-FOH-nee-kah*
Bill my credit card.	**L'addebiti alla mia carta di credito.** *lahd-DEH-bee-tee AHL-lah MEE-ah KAHR-tah dee KREH-dee-toh*
May I bill the charges to my room?	**Posso addebitare i costi alla mia stanza?** *POHS-soh ahd-deh-bee-TAH-reh ee KOHS-tee AHL-lah MEE-ah STAHN-tsah*
Information, please.	**Informazioni, per favore.** *een-fohr-mah-TSYOH-nee pehr fah-VOH-reh*
I'd like the number for ____.	**Vorrei il numero per ____.** *vohr-RAY eel NOO-meh-roh pehr*
I just got disconnected.	**È caduta la linea.** *EH kah-DOO-tah lah LEE-neh-ah*
The line is busy.	**La linea è occupata.** *lah LEE-neh-ah EH ohk-koo-PAH-tah*

COMMUNICATIONS

INTERNET ACCESS

Where is an Internet café?

Dove si trova un punto di accesso a Internet?
DOH-veh see TROH-vah oon POON-toh di aht-CHESS-soh ah internet

Is there Wi-Fi?

C'è un collegamento Wi-Fi?
ch-EH oon kohl-leh-gah-MEHN-toh WI-Fi

How much do you charge per minute / hour?

Quanto costa al minuto / all'ora?
KWAHN-toh KOHS-tah ahl mee-NOO-toh / ahl-LOH-rah

Can I print here?

Posso stampare qui?
POHS-soh stahm-PAH-reh kwee

Can I burn a CD?

Posso copiare un CD?
POHS-soh koh-PYAH-reh oon chee-DEE

Would you please help me change the language preference to English?

Può aiutarmi a cambiare l'impostazione della lingua all'inglese?
PWOH ah-yoo-TAHR-mee ah kahm-BYAH-reh leem-pohs-tah-TSYOH-neh DEHL-lah LEEN-gwah ahl-leen-GLEH-seh

May I scan something?	**Posso scannerizzare qualcosa?** *POHS-soh skahn-neh-reed-DZAH-reh kwahl-KOH-zah*
Can I upload photos?	**Posso caricare foto?** *POHS-soh kah-ree-KAH-reh PHOH-toh*
Do you have a Mac?	**C'è un Mac?** *ch-EH oon mac*
Do you have a PC?	**C'è un PC?** *ch-EH oon pee-CHEE*
Do you have a newer version of this software?	**C'è una versione più recente di questo software?** *ch-EH OO-nah vehr-SYOH-neh PYOO reh-CHEHN-teh dee KWEHS-toh software*
Do you have broadband?	**C'è il broadband?** *ch-EH eel broadband*
How fast is your connection speed?	**Che velocità di connessione c'è qui?** *keh veh-loh-chee-TAH dee kohn-nehs-SYOH-neh ch-EH kwee*

GETTING MAIL

Where is the post office?	**Dov'è l'ufficio postale?** *doh-VEH loof-FEE-choh pohs-TAH-leh*
May I send an international package?	**Posso inviare un pacco internazionale?** *POHS-soh een-VYAH-reh oon PAHK-koh een-tehr-nah-tsyoh-NAH-leh*
Do I need a customs form?	**Occorre il modulo doganale?** *ohk-KOHR-reh eel MOH-doo-loh doh-gah-NAH-leh*
Do you sell insurance for packages?	**Offrite l'assicurazione per i pacchi?** *ohf-FREE-teh lahs-see-koo-raht-SYOH-neh pehr ee PAHK-kee*
Please, mark it fragile.	**Lo contrassegni come fragile, per favore.** *loh kohn-trahs-SEHN-njee KOH-meh FRAH-jee-leh pehr fah-VOH-reh*
Please, handle with care.	**Lo maneggi con cura, per favore.** *loh mah-NEHD-jee kohn KOO-rah pehr fah-VOH-reh*
Do you have twine?	**Ha dello spago?** *AH DEHL-loh SPAH-goh*
Do you have a twine clamp? (sometimes required for packages)	**Ha un sigillo?** *AH oon see-GEEL-loh*
Where is a DHL (express) office?	**Dov'è un ufficio DHL?** *doh-VEH oon oof-FEE-choh dee-akka-EHLLEH*

Listen Up: Postal Lingo

Il prossimo!
eel PROHS-see-moh

Next!

Lo metta qui.
loh MEHT-tah kwee

Set it here.

Come lo vuole inviare?
*KOH-meh loh VWOH-leh
een-VYAH-reh*

How would you like to send it?

**Che tipo di servizio
desidera?**
*keh TEE-poh dee sehr-
VEET-syoh deh-ZEE-deh-rah*

What kind of service would
you like?

Come posso aiutarla?
*KOH-meh POHS-soh
ah-yoo-TAHR-lah*

How can I help you?

Consegne
kohn-SEHN-nyeh

Dropoff window

Ritiri
ree-TEE-ree

Pickup window

Do you sell stamps?	**Vende francobolli?** *VEHN-deh frahn-koh-BOHL-lee*
Do you sell postcards?	**Vende cartoline?** *VEHN-deh kahr-toh-LEE-neh*
May I send that first class?	**Posso inviarlo con la Posta prioritaria?** *POHS-soh een-VYAHR-loh kohn lah POHS-tah pree-oh-ree-TAH-ryah*
How much to send that express / air mail?	**Quanto costa inviarlo come pacco espresso? / per posta aerea?** *KWAHN-toh KOHS-tah KOH-meh PAHK-koh eh-SPREH-soh / pehr POHS-tah ah-EH-reh-ah*

Do you offer overnight delivery?

Offrite la consegna il giorno dopo?
ohf-FREE-teh lah kohn-SEHN-nyah eel JOHR-noh DOH-poh

How long will it take to reach the United States?

Quanto tempo ci vorrà per arrivare negli Stati Uniti?
KWAHN-toh TEHM-poh chee vohr-RAH pehr ahr-ree-VAH-reh NEL-lyee STAH-tee oo-NEE-tee

I'd like to buy an envelope.

Vorrei comprare una busta.
vohr-RAY kohm-PRAH-reh OO-nah BOOS-tah

May I send it airmail?

Posso inviarlo per posta aerea?
POHS-soh een-VYAHR-loh pehr POHS-tah ah-EH-reh-ah

I'd like to send it certified / registered mail.

Vorrei inviarlo per posta raccomandata.
vohr-RAY een-VYAHR-loh pehr POHS-tah rahk-koh-mahn-DAH-tah

CHAPTER SEVEN

CULTURE

CINEMA

Is there a movie theater nearby?	**C'è un cinema qui vicino?** *ch-EH oon CHEE-neh-mah kwee vee-CHEE-noh*
What's playing tonight?	**Cosa danno stasera?** *KOH-zah DAHN-noh stah-SEH-rah*
Is that in English or Italian?	**È in inglese o in italiano?** *EH een een-GLEH-seh oh een ee-tah-LYAH-noh*
Are there English subtitles?	**Ci sono sottotitoli in inglese?** *chee SOH-noh soht-toh-TEE-toh-lee een een-GLEH-seh*
Is the theater air conditioned?	**La sala è climatizzata?** *lah SAH-lah EH klee-mah-teed-ZAH-tah*
How much is a ticket?	**Quanto costa un biglietto?** *KWAHN-toh KOHS-tah oon beel-LYEHT-toh*
Do you have _____	**Ci sono sconti per _____** *chee SOH-noh SKOHN-tee pehr*
senior discounts?	**anziani?** *ahn-TSYAH-nee*
student discounts?	**studenti?** *stoo-DEHN-tee*
children discounts?	**bambini?** *bahm-BEE-nee*

What time is the movie showing?	**A che ora comincia lo spettacolo?** *ah keh OH-rah koh-MEEN-chah loh speht-TAH-koh-loh*
How long is the movie?	**Quanto dura il film?** *KWAHN-toh DOO-rah eel film*
May I buy tickets in advance?	**Posso comprare i biglietti in anticipo?** *POHS-soh kohm-PRAH-reh ee beel-LYEHT-tee een ahn-TEE-chee-poh*
Is it sold out?	**È tutto esaurito?** *EH TOOT-toh eh-zow-REE-toh*
When does it begin?	**Quando inizia?** *KWAHN-doh ee-NEE-tsyah*

PERFORMANCES

Are there any plays showing right now?	**Ci sono spettacoli teatrali al momento?** *chee SOH-noh speht-TAH-koh-lee tehah-TRAH-lee ahl moh-MEHN-toh*
Where can I buy tickets?	**Dove posso comprare i biglietti?** *DOH-veh POHS-soh kohm-PRAH- reh ee beel-LYEHT-tee*
Are there student discounts?	**Ci sono sconti per studenti?** *chee SOH-noh SKOHN-tee pehr stoo-DEHN-tee*
I need ____ seats.	**Mi servono ____ posti.** *mee SEHR-voh-noh ____ POHS-tee*

For a full list of numbers, see p7.

CULTURE

Listen Up: Box Office Lingo

Cosa le piacerebbe vedere?   What would you like to see?
KOH-zah leh pyah-cheh-
REHB-beh veh-DEH-reh

Quanti?   How many?
KWAHN-tee

Per due adulti?   For two adults?
pehr DOO-eh ah-DOOL-tee

Vuole dei popcorn?   Would you like some popcorn?
VWOH-leh DEH-ee popcorn

Vuole altro?   Would you like anything else?
VWOH-leh AHL-troh

An aisle seat, please.   **Un posto sul corridoio, per favore.**
oon POHS-toh sool kohr-ree-
DOH-yoh pehr fah-VOH-reh

An orchestra seat.   **Un posto in platea.**
oon POHS-toh een plah-TEH-ah

What time does the play   **A che ora inizia lo spettacolo?**
start?   *ah keh OH-rah ee-NEE-tsyah loh*
speht-TAH-koh-loh

Is there an intermission?   **C'è un intervallo?**
ch-EH oon een-tehr-VAHL-loh

Do you have an opera   **C'è un teatro dell'opera?**
house?   *ch-EH oon teh-AH-troh*
dehl-LOH-peh-rah

Is there a local symphony?	**C'è un'orchestra locale?** *ch-EH oon-ohr-KEHS-trah loh-KAH-leh*
May I purchase tickets over the phone / online?	**Posso comprare i biglietti per telefono / online?** *POHS-soh kohm-PRAH-reh ee beel-LYEHT-tee pehr teh-LEH-foh-noh / on-line*
What time is the box office open?	**A che ora apre il botteghino?** *ah keh OH-rah AH-preh eel boht-teh-GHEE-noh*
I need space for a wheelchair, please.	**Ho bisogno di spazio per una sedia a rotelle, per favore.** *OH bee-ZOHN-nyoh dee SPAH-tsyoh pehr OO-nah SEH-dyah ah roh-TEHL-leh pehr fah-VOH-reh*
Do you have private boxes available?	**Sono disponibili dei palchi privati?** *SOH-noh dees-poh-NEE-bee-lee DEH-ee PAHL-kee pree-VAH-tee*
Is there a church that gives concerts?	**C'è una chiesa che dà concerti?** *ch-EH OO-nah KYEH-zah keh DAH kohn-CHEHR-tee*
A program, please.	**Un programma, per favore.** *oon proh-GRAHM-mah pehr fah-VOH-reh*
Please show us our seats.	**Ci può mostrare i nostri posti, per favore.** *chee PWOH MOHS-trah-reh ee NOSS-treeh POHS-tee pehr fah-VOH-reh*

CULTURE

MUSEUMS, GALLERIES, AND SIGHTS

Do you have a museum guide?	**Ha una guida al museo?** *AH OO-nah GWEE-dah ahl moo-ZEH-oh*
Do you have guided tours?	**Ci sono visite guidate?** *chee SOH-noh VEE-zee-teh gwee-DAH-teh*
What are the museum hours?	**Quali sono gli orari del museo?** *KWAH-lee SOH-noh lyee oh-RAH-ree dehl moo-ZEH-oh*
Do I need an appointment?	**Serve un appuntamento?** *SEHR-veh oon ahp-poon-tah-MEHN-toh*
What is the admission fee?	**Quanto costa l'ingresso?** *KWAHN-toh KOHS-tah leen-GREHS-soh*
Do you have discounts for ___	**Ci sono sconti per___** *chee SOH-noh SKOHN-tee pehr*
students?	**studenti?** *stoo-DEHN-tee*
seniors?	**anziani?** *ahn-TSYAH-nee*
children?	**bambini?** *bahm-BEE-nee*
Do you have services for the hearing impaired?	**Ci sono servizi per ipoudenti?** *chee SOH-noh sehr-VEET-see pehr ee-poh-oo-DEHN-tee*
Do you have audio tours in English?	**Ci sono audioguide in inglese?** *chee SOH-noh ahoo-dyo-GWEE-deh een een-GLEH-seh*

CHAPTER EIGHT

SHOPPING

This chapter covers the phrases you'll need to shop in a variety of settings: from the mall to the town square artisan market. We also threw in the terminology for a visit to the barber or hairdresser.

For coverage of food and grocery shopping, see Chapter Four, Dining.

GENERAL SHOPPING TERMS

Please tell me _____	**Può dirmi per favore _____** *PWOH DEER-mee pehr* *fah-VOH-reh*
how to get to a mall?	**come si arriva ad un centro commerciale?** *KOH-meh see ahr-REE-vah ahd oon CHEHN-troh kohm-mehr-CHAH-leh*
the best place for shopping?	**il posto migliore per fare compere?** *eel POHS-toh meel-LYOH-reh pehr FAH-reh KOHM-peh-reh*
how to get downtown?	**come si arriva in centro?** *KOH-meh see ahr-REE-vah een CHEHN-troh*

Closed for August

Cities grow hot and steamy in summer, so Italians head for the hills—or the beach. Family-run restaurants and some urban tourist attractions begin closing on August 1st. The exodus peaks on August 15th, Ferragosto *fehr-rah-GOHS-toh* (Assumption Day), which celebrates the Virgin Mary's ascent to heaven.

Where can I find a _____	**Dove posso trovare _____**
	DOH-veh POHS-soh troh-VAH-reh
shoe store?	**un negozio di scarpe?**
	oon neh-GOH-tsyoh dee SKAHR-peh
clothing store for men / women / children?	**un negozio di abbigliamento per uomo / donna / bambini?**
	oon neh-GOH-tsyoh dee ahb-beel-lyah-MEHN-toh pehr WOH-moh / DOHN-nah / bahm-BEE-nee
designer fashion shop?	**una boutique?**
	OO-nah boutique
vintage clothing store?	**un negozio di abiti usati?**
	oon neh-GOH-tsyoh dee AH-bee-tee oo-ZAH-tee
jewelry store?	**una gioielleria?**
	OO-nah joh-yehl-leh-REE-ah
bookstore?	**una libreria?**
	OO-nah lee-breh-REE-ah
toy store?	**un negozio di giocattoli?**
	oon neh-GOH-tsyoh dee joh-KAHT-toh-lee
stationery store?	**una cartoleria?**
	OO-nah kahr-toh-leh-REE-ah
antiques shop?	**un negozio di antichità?**
	oon neh-GOH-tsyoh dee ahn-tee-kee-TAH
cigar shop?	**un tabaccaio?**
	oon tah-bahk-KAH-yoh
souvenir shop?	**un negozio di souvenir?**
	oon neh-GOH-tsyoh dee souvenir
Where can I find a flea market?	**Dove posso trovare un mercatino?**
	DOH-veh POHS-soh troh-VAH-reh oon mehr-kah-TEE-noh

CLOTHES SHOPPING

I'd like to buy _____

Vorrei comprare _____
vohr-RAY kohm-PRAH-reh

men's shirts.

delle camicie da uomo.
DEHL-leh kah-MEE-cheh dah WOH-moh

women's shoes.

delle scarpe da donna.
DEHL-leh SKAHR-peh dah DOHN-nah

children's clothes.

dei vestiti per bambini.
day vehs-TEE-tee pehr bahm-BEE-nee

For a full listing of numbers, see p7.

I'm looking for a size _____

Cerco una taglia _____
CHEHR-koh OO-nah TAHL-lyah

extra-small.

molto piccola.
MOHL-toh PEEK-koh-lah

small.

piccola.
PEEK-koh-lah

medium.

media.
MEH-dyah

large.

grande.
GRAHN-deh

extra-large.

molto grande.
MOHL-toh GRAHN-deh

I'm looking for _____

Cerco _____
CHEHR-koh

a silk blouse.

una camicia di seta.
OO-nah kah-MEE-chah dee SEH-tah

cotton pants.

dei pantaloni di cotone.
day pahn-tah-LOH-nee dee koh-TOH-neh

a hat.

un cappello / berretto.
oon kahp-PEHL-loh / behr-REHT-toh

Gli orecchini

La collana

Il vestito

L'orologio

I tacchi alti

La camicia

La cravatta

La giacca

La cintura

I pantaloni

Le scarpe

sunglasses.	**degli occhiali da sole.**
	DEHL-lyee OHK-kyah-lee dah SOH-leh
underwear.	**della biancheria intima.**
	DEHL-lah byahn-keh-REE-ah EEN-tee-mah
cashmere.	**qualcosa in cashmere.**
	kwahl-KOH-zah een cashmere
socks.	**dei calzini.**
	day kahl-TSEE-nee
sweaters.	**delle maglie.**
	DEHL-leh MAHL-lyeh
a coat.	**una giacca.**
	OO-nah JAHK-kah
a swimsuit.	**un costume da bagno.**
	oon kohs-TOO-meh dah BAHN-nyoh

gli ochialli

la maglietta

i jeans

le scarpe da tennis

May I try it on?	**Posso provarlo?** *POHS-soh proh-VAHR-loh*
Where can I try this on?	**Dove posso provarlo?** *DOH-veh POHS-soh proh-VAHR-loh*
This is _____	**Questo è _____** *KWEHS-toh EH*
too tight.	**troppo stretto.** *TROHP-poh STREHT-toh*
too loose.	**troppo largo.** *TROHP-poh LAHR-goh*
too long.	**troppo lungo.** *TROHP-poh LOON-goh*
too short.	**troppo corto.** *TROHP-poh KOHR-toh*
This fits great!	**È perfetto!** *EH pehr-FEHT-toh*
Thanks, I'll take it.	**Grazie, lo prendo.** *GRAH-tsyeh loh PREHN-doh*

Do you have that in ___	**Ha ___**
	AH
a smaller / larger size?	**una taglia più piccola / grande?**
	OO-nah TAHL-lya PYOO
	PEEK-koh-lah / GRAHN-deh
a different color?	**un altro colore?**
	oon AHL-troh koh-LOH-reh
How much is it?	**Quanto costa?**
	KWAHN-toh KOHS-tah

ARTISAN MARKET SHOPPING

Is there a craft / artisan market?	**C'è un mercato di artigianato?**
	ch-EH oon mehr-KAH-toh dee
	ahr-tee-jah-NAH-toh
That's beautiful. May I look at it?	**Che bello. Posso vederlo?**
	keh BEHL-loh POHS-soh
	veh-DEHR-loh
When is the farmers' market open?	**Quando apre il mercato di frutta e verdura?**
	KWAHN-doh AH-preh eel
	mehr-KAH-toh dee FROOT-tah eh
	vehr-DOO-rah

For full coverage of time, see p12.

Is that open every day of the week?	**È aperto tutti i giorni della settimana?**
	EH ah-PEHR-toh TOOT-tee ee
	JOHR-nee DEHL-lah
	seht-tee-MAH-nah

For full coverage of days of the week, see p14.

How much does that cost?	**Quanto costa?**
	KWAHN-toh KOHS-tah
That's too expensive.	**È troppo caro.**
	EH TROHP-poh KAH-roh

Venditori di falsi (fake designer goods)

Beware of unscrupulous vendors who attempt to sell you illegal, contraband, or fake goods. Recent laws penalize buyers as well as sellers.

How much for two?

A quanto me lo mette se ne prendo due?
ah KWAHN-toh meh loh MEHT-teh seh neh PREHN-doh DOO-eh

Do I get a discount if I buy two or more?

Mi fa lo sconto se ne compro due o più?
mee fah loh SKOHN-toh seh neh KOHM-proh DOO-eh oh PYOO

Do I get a discount if I pay in cash?

Mi fa lo sconto se pago in contanti?
mee fah loh SKOHN-toh seh PAH-goh een kohn-TAHN-tee

No thanks. Maybe I'll come back.

No grazie. Magari torno.
noh GRAH-tsyeh mah-GAH-ree TOHR-noh

Would you take € _____?

Vanno bene ___ euro?
VAHN-noh BEH-neh ___ EH-oo-roh

For a full list of numbers, see p7.
That's a deal!

Affare fatto!
ahf-FAH-reh FAHT-toh

Do you have a less expensive one?
Is there VAT tax?

Ne ha uno meno caro?
neh AH OO-noh MEH-no KAH-roh
C'è l'IVA?
ch-EH LEE-vah

May I have the VAT forms?

Posso avere un modulo per il rimborso dell'IVA?
POHS-soh ah-VEH-reh oon MOH-doo-loh pehr eel reem-BOHR-soh dehl-LEE-vah

BOOKSTORE / NEWSSTAND SHOPPING

Is there a ____ nearby?

C'è ____ qui vicino?
ch-EH ____ kwee vee-CHEE-noh

bookstore

una libreria
OO-nah lee-breh-REE-ah

newsstand

un'edicola
oon-eh-DEE-koh-lah

Do you have ____ in English?

Avete ____ in inglese?
ah-VEH-teh ____ een een-GLEH-seh

books

libri
LEE-bree

newspapers

giornali
johr-NAH-lee

magazines

riviste
ree-VEES-teh

books about local history

libri di storia locale
LEE-bree dee STOH-ryah loh-KAH-leh

picture books

libri illustrati
LEE-bree eel-loos-TRAH-tee

travel guides

guide turistiche
GWEE-deh too-REES-tee-keh

maps

cartine
kahr-TEE-neh

SHOPPING FOR ELECTRONICS

With some exceptions, shopping for electronic goods in Italy is generally not recommended for North Americans. The PAL encoding system for TVs is different from NTSC and will not work in the United States or Canada. Most DVDs are region encoded to work in Europe only.

Can I play this in the U.S.?	**Funziona questo negli Stati Uniti?** *foon-TSYOH-nah KWEHS-toh NEHL-lyee STAH-tee oo-NEE-tee*
Will this game work on my game console in the U.S.?	**Questo gioco funziona su una console americana?** *KWEHS-toh JOH-koh foon-TSYOH-nah soo OO-nah kohn-SOHL ah-meh-ree-KAH-nah*
Is this DVD region encoded?	**Questo DVD possiede vari codici regionali?** *KWES-toh DVD pohs-SYEH-deh ee VAH-ree COH-dee-chee reh-djoh-NAH-leh*
Will this work with a 110V AC adaptor?	**Questo funziona con un adattatore da 110 volts?** *KWEHS-toh foon-TSYOH-nah kohn oon ah-daht-tah-TOH-reh dah CHEHN-toh-DYEH-chee volts*
Do you have an adaptor plug for 110 to 220 volts?	**Avete un adattatore da 110 a 220 volts?** *Ah-VEH-teh oon ah-daht-tah-TOH-reh dah CHEHN-toh-DYEH-chee ah doo-eh-CHEHN-toh-VEHN-tee volts*
Do you sell electronic adaptors here?	**Vendete adattatori di corrente?** *vehn-DEH-teh ah-dah-tah-TOH-ree dee kohr-REHN-teh*

Is it safe to use my laptop with this adaptor?	**Posso usare il computer portatile con questo adattatore?** *POHS-soh oo-ZAH-reh eel com puter pohr-TAH-tee-leh kohn KWEHS-toh ah-daht-tah-TOH-reh*
If it doesn't work, may I return it?	**Se non funziona, posso portarlo indietro?** *seh nohn foon-TSYOH-nah POHS-soh pohr-TAHR-loh een-DYEH-troh*
May I try it here in the store?	**Posso provarlo qui in negozio?** *POHS-soh proh-VAHR-loh kwee een neh-GOH-tsyoh*

AT THE BARBER / HAIRDRESSER

Do you have a style guide?	**Ha un catalogo delle pettinature?** *AH oon kah-TAH-loh-goh DEHL-leh peht-tee-nah-TOO-reh*
A trim, please.	**Una spuntatina, per favore.** *OO-nah spoon-tah-TEE-nah pehr fah-VOH-reh*
I'd like it bleached.	**Vorrei ossigenarli.** *vohr-RAY ohs-see-jeh-NAHR-lee*
Would you change the color _____	**Mi fa il colore _____** *mee fah eel koh-LOH-reh*
darker?	**più scuro?** *PYOO SKOO-roh*
lighter?	**più chiaro?** *PYOO KYAH-roh*

For a full list of personal descriptions, see p117.
For a full list of colors, see English / Italian dictionary.

Would you just touch it up a little?

Me li sistema un po'?
meh lee see-STEH-mah oon POH

I'd like it curled.

Vorrei farmi i ricci.
vohr-RAY FAHR-mee ee REE-chee

Do I need an appointment?

Devo prendere un appuntamento?
DEH-voh prehn-DEH-reh oon ahp-poon-tah-MEHN-toh

May I make an appointment?

Posso prendere un appuntamento?
POHS-soh PREHN-deh-reh oon ahp-poon-tah-MEHN-toh

Do you do permanents?

Fate permanenti?
FAH-teh pehr-mah-NEHN-tee

Please use low heat.

Usi il phon a bassa temperatura, per favore.
OO-zee eel FOHN ah BAHS-sah tehm-peh-rah-TOO-rah pehr fah-VOH-reh

Please don't blow dry it.

Non li asciughi col phon, per favore.
nohn lee ah-SHOOG-ee kohl fohn pehr fah-VOH-reh

Please dry it curly / straight

Li asciughi ricci / lisci, per favore.
lee ah-SHOO-gy REE-chee / LEE-shee pehr fah-VOH-reh

Would you fix my highlights?

Mi fa i colpi di sole?
mee fah ee KOHL-pee dee SOH-leh

Do you wax?

Fate la ceretta?
FAH-teh lah cheh-REHT-tah

I'd like a Brazilian wax.

Vorrei una ceretta brasiliana.
vohr-REHY OO-nah cheh-REHT-tah brah-zee-LYA-nah

Please wax my ___

Mi faccia la ceretta ___ per favore.
mee FAHT-chah lah cheh-REHT-tah ___ pehr fah-VOH-reh

legs.

alle gambe
AHL-leh GAHM-beh

bikini line.

alla zona bikini
AHL-lah DZOH-nah bikini

eyebrows.

alle sopracciglia
AHL-leh soh-praht-CHEEL-lyah

under my nose.

sotto il naso
SOHT-toh eel NAH-zoh

Please trim my beard.

Mi spunti la barba, per favore.
mee SPOON-tee lah BAHR-bah pehr fah-VOH-reh

A shave, please.

Mi faccia la barba, per favore.
mee FAHT-chah lah BAHR-bah pehr fah-VOH-reh

Use a fresh blade, please.

Usi una lametta nuova, per favore.
OO-zee OO-nah lah-MEHT-tah NWOH-vah pehr fah-VOH-reh

Sure, cut it all off.

Certo, la tagli tutta.
CHEHR-toh lah TAHL-lyee TOOT-tah

CHAPTER NINE
SPORTS & FITNESS

STAYING FIT

Is there a gym nearby?

C'è una palestra qui vicino?
ch-EH OO-nah pah-LEHS-trah kwee vee-CHEE-noh

Does the hotel have a gym?

C'è una palestra nell'hotel?
cheh OO-nah pah-LEH-strah nehl-loh-TEHL

Do you have free weights?

Avete pesi liberi?
ah-VEH-teh PEH-zee LEE-beh-ree

Is there a pool?

C'è una piscina?
cheh OO-nah pee-SHEE-nah

Do I have to be a member?

Devo essere socio?
DEH-voh EHS-seh-reh SOH-choh

May I come here for one day?

Posso venire qui per un giorno?
POHS-soh veh-NEE-reh kwee pehr oon JOHR-noh

How much does a membership cost?

Quanto costa iscriversi?
KWAHN-toh KOHS-tah ee-SCREE-vehr-see

I need to get a locker, please.	**Mi serve un armadietto, per favore.**
	mee SEHR-veh oon ahr-mah-DYEHT-toh pehr fah-VOH-reh
Do you have locks?	**Avete dei lucchetti?**
	Ah-VEH-teh day look-KEHT-tee
Do you have _____	**Avete _____**
	ah-VEH-teh
a treadmill?	**il tapis roulant?**
	eel tah-PEE-roo-lahn
a stationary bike?	**la cyclette?**
	lah see-KLEHT
handball / squash courts?	**campi da pallamano / squash?**
	KAHM-pee dah PAHL-lah-MAH-noh / squash
Are they indoors?	**I campi sono al coperto?**
	ee KAHM-pee SOH-noh ahl koh-PEHR-toh
I'd like to play tennis.	**Vorrei giocare a tennis.**
	vohr-RAY joh-KAH-reh ah tennis
Would you like to play?	**Vuole giocare?**
	VWOH-leh joh-KAH-reh
I'd like to rent a racquet.	**Vorrei noleggiare una racchetta.**
	vohr-RAY noh-lehd-JAH-reh OO-nah rahk-KEHT-tah
I need to buy some _____	**Devo comprare _____**
	DEH-voh kohm-PRAH-reh
new balls.	**delle palle nuove.**
	DEHL-leh PAHL-leh NWOH-veh
safety glasses.	**degli occhiali protettivi.**
	DEHL-lyee ohk-KYAH-lee proh-teht-TEE-vee
May I reserve a court for tomorrow?	**Posso prenotare un campo per domani?**
	POHS-soh preh-noh-TAH-reh oon KAHM-poh pehr doh-MAH-nee

May I have clean towels?	**Posso avere degli asciugamani puliti?** *POHS-soh ah-VEH-reh dehl-lyee ah-shoo-gah-MAH-nee poo-LEE-teh*
Where are the showers / locker-rooms?	**Dove sono le docce / gli spogliatoi?** *DOH-veh SOH-noh leh DOHT-cheh l lyee spohl-lyah-TOY*
Do you have a workout room for women only?	**Avete una sala riservata alle donne?** *ah-VEH-teh OO-nah SAH-lah ree-sehr-VAH-tah AHL-leh DOHN-neh*
Do you have aerobics classes?	**Avete corsi di aerobica?** *Ah-VEH-teh KOHR-see dee ah-eh-ROH-bee-kah*
Are there Yoga / Pilates classes?	**Ci sono lezioni di yoga / pilates?** *chee SOH-noh leh-TZYOH-nee dee YO-ga l pee-LAH-tehs*
Do you have a women's pool?	**Avete una piscina per donne?** *Ah-VEH-teh OO-nah pee-SHEE-nah pehr DOHN-neh*
Let's go for a jog.	**Andiamo a fare jogging.** *anh-DYAH-moh ah FAH-reh jogging*
That was a great workout!	**Che bell'allenamento!** *keh BEHL-lahl-leh-nah-MEHN-toh*

CATCHING A GAME

Where is the stadium?	**Dov'è lo stadio?** *doh-VEH loh STAH-dyoh*
Who's playing?	**Chi gioca?** *kee JYO-kah*

Who is the best goalie?	**Chi è il portiere più bravo?** *Kee-LEH eel pohr-TYEH-reh* *PYOO BRAH-voh*
Do you have any amateur / professional teams?	**Ci sono squadre dilettanti / professioniste?** *chee SOH-noh SKWAH-dreh* *dee-leht-TAHN-tee /* *proh-fehs-syoh-NEES-teh*
Is there a game I could play in?	**C'è una partita in cui posso partecipare?** *ch-EH OO-nah paar-TEE-tah een* *KOO-ee POHS-soh* *pahr-teh-chee-PAH-reh*
Which is the best team?	**Qual è la squadra migliore?** *kwah-LEH lah SKWAH-drah* *meel-LYOH-reh*
Will the game be on television?	**Questa partita sarà trasmessa in TV?** *KWEHS-tah paar-TEE-tah sah-RAH* *trahs-MEHS-sah een TV*
Where can I buy tickets?	**Dove si comprano i biglietti?** *DOH-veh see KOHM-prah-noh ee* *beel-LYEHT-tee*
The best seats, please.	**I posti migliori, per favore.** *ee POHS-tee meel-LYOH-ree pehr* *fah-VOH-reh*
The cheapest seats, please.	**I posti più economici, per favore.** *ee POHS-tee PYOO eh-koh-NOH-* *mee-chee pehr fah-VOH-reh*
How close are these seats?	**Quanto sono lontani questi posti?** *KWAHN-toh SOH-noh lohn-TAH-* *nee KWEHS-tee POHS-tee*
Where are these seats?	**Dove sono questi posti?** *DOH-veh SOH-noh KWES-tee* *POHS-tee*

SPORTS & FITNESS

May I have box seats?	**Posso avere dei posti in tribuna?**
	POHS-soh ah-VEH-reh day POHS-tee een TREE-boo-nah
Wow! What a game!	**Wow! Che partita!**
	wow keh paar-TEE-tah
Go! Go! Go!	**Vai, vai, vai! / Dai, dai, dai!**
	VAH-ee / DAH-ee
Go for it!	**Forza! Vai!**
	FOHR-tsah VAH-ee
Score!	**Gol!**
	gol
What's the score?	**Qual è il punteggio?**
	kwah-LEH eel poon-TEHD-joh
Who's winning?	**Chi sta vincendo?**
	kee stah veen-CHEHN-doh

HIKING

Is there a trail map?	**C'è una cartina dei sentieri?**
	CEH OO-nah kahr-TEE-nah DEH-ee sehn-TYE-ree
Do we need to hire a guide?	**Dobbiamo ingaggiare una guida?**
	dohb-BYAH-moh neen-gah-DJAH-ray OO-nah GWEE-dah
Where can I rent equipment?	**Dove si può noleggiare l'attrezzatura?**
	DOH-veh see pwoh noh-lehd-JAH-reh lah-treh-TSAH-too-rah
Do they have rock climbing there?	**Si possono fare arrampicate qui?**
	see POHS-soh-noh FAH-reh ahr-rahm-pee-KAH-teh kwee
We need more ropes and carabiners.	**Ci servono altre corde e moschettoni.**
	chee SEHR-voh-noh AHL-treh KOHR-deh eh mohs-keht-TOH-nee

Where can we go mountain climbing?	**Dove si può andare a fare scalate?** *DOH-veh see PWOH ahn-DAH-reh ah FAH-reh skah-LAH-teh*
Are the routes _____	**I sentieri sono _____** *ee sehn-TYEH-ree SOH-noh*
well marked?	**ben marcati?** *behn mahr-KAH-tee*
in good condition?	**in buone condizioni?** *een BWOH-neh kohn-dee-TSYOH-nee*
What is the altitude there?	**Che altitudine c'è là?** *keh ahl-tee-TOO-dee-neh ch-EHLAH*
How long will it take?	**Quanto tempo ci vorrà?** *KWAHN-toh TEHM-poh chee vohr-RAH*
Is it very difficult?	**È molto difficile?** *EH MOHL-toh deef-FEE-chee-leh*
Is there a fixed-protection climbing path?	**C'è una via ferrata?** *CHEH OO-nah VEE-ah fehr-RAH-tah*
I want to hire someone to carry my excess gear.	**Vorrei noleggiare un aiuto che mi porti l'attrezzatura extra.** *VOHR-ray noh-lehd-JAH-reh oon ah-YOO-toh keh mee POHR-tee laht-treht-tsah-TOO-rah extra*

We don't have time for a long route.	**Non c'è tempo per un percorso lungo.** *nohn ch-EH TEHM-poh pehr- oon pehr-KOHR-soh LOON-goh*
I don't think it's safe to proceed.	**Non mi pare sicuro avanzare.** *nohn mee PAH-reh see-KOO-roh ah-vahn-TSAH-reh*
Do we have a backup plan?	**Abbiamo un piano alternativo?** *ahb-BYAH-moh oon PYAH-noh ahl-tehr-nah-TEE-voh*
If we're not back by tomorrow, send a search party.	**Se non torniamo entro domani, mandate una squadra di ricerca.** *seh nohn tohr-NYAH-moh EHN-troh doh-MAH-nee mahn-DAH-teh OO-nah SKWAH-drah dee ree-CHEHR-kah*
Are the campsites marked?	**I siti per il campeggio sono marcati?** *ee SEE-tee pehr eel kahm-PEHD-joh SOH-noh mahr-KAH-tee*
Can we camp off the trail?	**Possiamo campeggiare fuori dal sentiero?** *pohs-SYAH-moh kahm-pehd-JAH-reh FWOH-ree dahl sehn-TYEH-roh*
Is it okay to build fires here?	**Si può accendere il fuoco qui?** *see pwoh ahc-CHEHN-deh-reh eel FWOH-koh kwee*
Do we need permits?	**Ci servono permessi?** *chee SEHR-voh-noh pehr-MEHS-see*

For more camping terms, see p79.

BOATING OR FISHING

I'd like to go fishing.

Mi piacerebbe andare a pesca.
mee pyah-ceh-REHB-beh ahn-DAH-reh ah PEH-skah

When do we sail?

Quando salpiamo?
KWAHN-doh sahl-PYAH-moh

Where are the life preservers?

Dove sono i salvagenti?
DOH-veh SOH-noh ee sahl-vah-JEHN-tee

Can I purchase bait?

Posso acquistare delle esche?
POHS-soh ah-kwees-TAH-reh DEHL-leh EHS-keh

Can I rent a rod / pole?

Posso noleggiare una canna da pesca?
POHS-soh noh-lehd-JAH-reh OO-nah KAHN-nah dah PEHS-kah

How long is the trip?

Quanto dura l'escursione?
KWAHN-toh DOO-rah lehs-koor-SYOH-neh

Are we going up river or down?

Andiamo a monte o a valle del fiume?
ahn-DYAH-moh ah MOHN-teh oh ah VAHL-leh dehl FYOO-meh

How far out are we going?

Quanto al largo andiamo?
KWAHN-toh ahl LAHR-goh ahn-DYAH-moh

How deep is the water here?

Quanto è profonda l'acqua qui?
KWAHN-toh EH proh-FOHN-dah LAHK-wah kwee

I got one!

L'ho preso!
LOH PREH-zoh

I can't swim.	**Non so nuotare.** *nohn soh nwo-TAH-reh*
Help! Lifeguard!	**Aiuto! Bagnino!** *AYU-to bahn-NY-noh*
Can we go ashore?	**Possiamo sbarcare?** *pohs-SYAH-moh zbahr-KAH-reh*

DIVING

I'd like to go snorkeling.	**Vorrei fare snorkeling.** *vohr-RAY FAH-reh snorkeling*
I'd like to go scuba diving.	**Vorrei fare delle immersioni.** *vohr-RAY FAH-reh DEHL-leh eem-mehr-SYO-nee*
I have a NAUI / PADI certification.	**Ho il certificato NAUI / PADI.** *OH eel chehr-tee-fee-KAH-toh NAUI / PADI*
I need to rent gear.	**Devo noleggiare l'attrezzatura.** *DEH-voh noh-lehd-JAH-reh laht-treht-tsah-TOO-rah*
We'd like to see some shipwrecks, if we can.	**Vorremmo vedere dei relitti di navi se possibile.** *vohr-REHM-moh veh-DEH-reh day reh-LEET-tee dee NAH-vee seh pohs-SEE-bee-leh*
Are there any good reef dives?	**Si possono fare belle immersioni lungo la scogliera?** *see POHS-soh-noh FAH-reh BEHL-leh eem-mehr-SYOH-nee LOON-goh lah skohl-LYEH-rah*
I'd like to see a lot of sea-life.	**Vorrei vedere tanta fauna marina.** *vohr-RAY veh-DEH-reh TAHN-tah FOW-nah mah-REE-nah*
Are the currents strong?	**Le correnti sono forti?** *leh kohr-REHN-tee SOH-noh FOHR-tee*
How clear is the water?	**L'acqua è limpida?** *LAHK-wah EH LEEM-pee-dah*

I want / don't want to go with a group.

Voglio / Non voglio andare in gruppo.
VOHL-lyoh / nohn VOHL-lyoh ahn-DAH-reh een GROOP-poh

Can we charter our own boat?

Possiamo noleggiare una barca per uso privato?
pohs-SYAH-moh noh-lehd-JAH-reh OO-nah BAHR-kah pehr OO-zoh pree-VAH-toh

AT THE BEACH

I'd like to rent _____ for a day / half a day.

Vorrei noleggiare ___ per un giorno / mezza giornata.
vohr-RAY noh-led-JAH-reh ___ pehr oon JOHR-noh / MED-zah johr-NAH-tah

a chair

una sdraio
OO-nah ZDRAH-yoh

an umbrella

un ombrellone
oon ohm-brehl-LOH-neh

Is there space _____

C'è posto ___
CHEH POHS-toh

closer to the water?

più vicino all'acqua?
PYOO vee-CHEE-noh ahl-LAHK-wah

away from the disco music?

lontano dalla musica?
lohn-TAH-noh DAHL-lah MOO-zee-kah

Equipped Beaches

Equipped beaches are more common than free ones, and they will charge for chair and umbrella rental, and for the use of changing facilities and shower.

Do you have ____
Avete ____
ah-VEH-teh

a bar?
un bar?
oon bar

a restaurant?
un ristorante?
oon rees-toh-RAHN-teh

games?
dei giochi?
day JOH-kee

a lifeguard?
un bagnino?
oon bahn-NYEE-noh

a kid's club?
animazione per bambini?
ah-nee-mah-TSYOH-neh pehr bahm-BEE-nee

pedal boats for rent / hire?
pedalò a noleggio?
peh-dah-LOH ah noh-LED-joh

Is there a free beach nearby?
C'è una spiaggia libera qui vicino?
ch-EH OO-nah SPYAD-jah LEE-beh-rah kwee vee-CHEE-noh

How are the currents?
Come sono le correnti?
KOH-meh SOH-noh leh kohr-REHN-tee

I'd like to go windsurfing.
Vorrei fare windsurf.
vohr-RAY FAH-reh windsurf

Is there wakeboarding / kitesurfing?
Avete il wakeboard / kitesurf?
ah-VEH-teh eel wakeboard / kitesurf

Can I rent equipment?
Posso noleggiare attrezzatura?
POHS-soh noh-lehd-JAH-reh aht-treht-sah-TOO-rah

GOLFING

I'd like to reserve a tee-time.

Vorrei prenotare un tee time.
vohr-RAY preh-noh-TAH-reh oon tee time

Do we need to be members to play?

Dobbiamo essere soci per giocare?
dohb-BYAH-moh EHS-seh-reh SOH-chee pehr joh-KAH-reh

How many holes is your course?

Quante buche ha il vostro campo da golf?
KWAHN-teh BOO-keh AH eel VOHS-troh KAHM-poh dah golf

What is par for the course?

Qual è la norma per il campo?
kwah-LEH lah NOHR-mah pehr eel KAHM-poh

What is the dress code for players?

Quali sono le regole per l'abbigliamento dei giocatori?
KWAH-lee SOH-noh leh reh-GOH-leh pehr LAHB-beel-lyah-MEHN-toh DEH-ee jyo-kah-TOH-ree

I need to rent clubs.

Devo noleggiare delle mazze.
DEH-voh noh-lehd-JAH-reh DEHL-leh MAHT-seh

I need to purchase a sleeve of balls.

Devo acquistare una confezione di palline.
DEH-voh ah-kwees-TAH-reh OO-nah kohn-feh-TSYOH-neh dee pahl-LEE-neh

Do you require soft spikes?

Ci vogliono i soft spikes?
chee VOHL-lyoh-noh ee soft spikes

Do you have carts?	**Avete i golf carts?**
	ah-VEH-teh ee golf carts
I'd like to hire a caddy.	**Vorrei noleggiare un caddy.**
	vohr-RAY noh-lehd-JAH-reh oon caddy
Do you have a driving range?	**Avete un campo di pratica?**
	ah-VEH-teh oon CAHM-poh dee PRAH-tee-kah
How much are the greens fees?	**Quanto sono le green fees?**
	KWAHN-toh SOH-noh leh green fees
Can I book a lesson with the pro?	**Posso prenotare una lezione con un professionista?**
	POHS-soh preh-noh-TAH-reh OO-nah leh-TSYOH-neh kohn oon proh-fehs-syoh-NEES-tah
I need to have a club repaired.	**Devo far riparare una mazza.**
	DEH-voh fahr ree-pah-RAH-reh OO-nah MAHT-sah
Is the course dry?	**Il campo è asciutto?**
	eel KAHM-poh EH ah-SHOOT-toh
Are there any wildlife hazards?	**Gli animali selvatici rappresentano un pericolo?**
	llyh ah-nee-MAH-lee sehl-VAH-tee-chee rahp-preh-SEHN-tah-noh oon peh-REE-koh-loh
How many meters is the course?	**Quanti metri misura il campo?**
	KWAHN-tee MEH-tree mee-ZOO-rah eel KAHM-poh

CHAPTER TEN

NIGHTLIFE

For coverage of movies and cultural events, see p144, Chapter Seven, "Culture."

NIGHTCLUBBING

Where can I find_____	**Dove posso trovare _____**
	DOH-veh POHS-soh troh-VAH-reh
a good nightclub?	**un bel locale notturno / night-club?**
	oon behl loh-KAH-leh noht-TOOR-noh / night club
a club with a live band?	**un locale con musica dal vivo?**
	oon loh-KAH-leh kohn MOO-zee-kah dahl VEE-voh
a reggae club?	**un locale con musica reggae?**
	oon loh-KAH-leh kohn MOO-zee-kah reggae
a hip hop club?	**un locale con musica hip hop?**
	oon loh-KAH-leh kohn MOO-zee-kah hip hop
a techno club?	**un locale con musica techno?**
	oon loh-KAH-leh kohn MOO-zee-kah techno
a gay / lesbian club?	**un locale gay?**
	oon loh-KAH-leh gay
a club where I can dance?	**una discoteca?**
	OO-nah dees-koh-TEH-kah
a club with Italian music?	**un locale con musica italiana?**
	oon loh-KAH-leh kohn MOO-zee-kah ee-tah-LYAH-nah
the most popular club in town?	**il locale più frequentato in città?**
	eel loh-KAH-leh PYOO freh-kwehn-TAH-toh een cheet-TAH

a piano bar?	**un piano bar?**
	oon piano bar.
the most upscale club?	**il locale più di lusso?**
	eel loh-KAH-leh PYOO dee
	LOOS-soh
What's the hottest bar these days?	**Qual è il bar più in voga al momento?**
	kwah-LEH eel bar PYOO een
	VOH-gah ahl moh-MEHN-toh
What's the cover charge?	**Quant'è il coperto?**
	kwahn-TEH eel koh-PEHR-toh
Do I need a membership?	**Bisogna essere soci?**
	bee-ZOHN-nyah EHS-seh-reh
	SOH-chee
Do they have a dress code?	**Che abbigliamento è richiesto?**
	keh ahb-beel-lyah-MEHN-toh EH
	ree-KYEHS-toh
Is it expensive?	**È caro?**
	EH KAH-roh
What's the best time to go?	**A che ora è meglio andarci?**
	ah keh OH-rah EH MEHL-lyoh
	ahn-DAHR-chee
What kind of music do they play there?	**Che tipo di musica c'è?**
	keh TEE-poh dee MOO-zee-kah
	ch-EH
Is there an outside area where I can smoke?	**C'è un'area fumatori all'esterno?**
	CEH oo-NAH-reh-ahh foo-mah-TOH-
	ree ahl-leh-STEHR-noh

I'd like a pack of cigarettes.

Vorrei un pacchetto di sigarette.
vohr-RAY oon pahk-KEHT-toh dee see-gah-REHT-teh

I'd like _____, please.

Vorrei _____, per favore.
vohr-RAY pehr fah-VOH-reh

  a drink

  qualcosa da bere
  kwahl-KOH-zah dah BEH-reh

  a bottle of beer

  una bottiglia di birra
  OO-nah boht-TEEL-lyah dee BEER-rah

  a beer on tap

  una birra alla spina
  OO-nah BEER-rah AHL-lah SPEE-nah

  a shot of _____

  un _____
  oon _____

For a full list of drinks, see p88.

Make it a double, please!

Doppio, per favore!
DOHP-pyoh pehr fah-VOH-reh

With ice, please.

Con ghiaccio, per favore.
kohn GYAT-choh pehr fah-VOH-reh

And one for the lady / the gentleman!

E uno / a per la signora / il signore!
eh OO-noh pehr lah seen-NYOH rah / eel seen-NYOH reh

How much for a bottle / glass of beer?

Quanto costa la birra alla bottiglia / al bicchiere?
KWAHN-toh KOHS-tah lah BEER-rah AHL-lah boht-TEEL-lyah / ahl beek-KYEH-reh

I'd like to buy a drink for that girl / guy over there.

Vorrei offrire da bere a quella ragazza / quel ragazzo là.
vohr-RAY ohf-FREE-reh dah BEH-reh ah KWEHL-lah rah-GAHT-sah / kwehl rah-GAHT-soh LAH

May I run a tab?

Posso mettere sul conto?
POHS-soh meht-TEH-reh sool KOHN-toh

ACROSS A CROWDED ROOM

Excuse me, may I buy you a drink?

Scusami, posso offrirti qualcosa da da bere?
SKOO-zah-mee POHS-soh ohf-FREER-tee kwahl-KOH-zah dah BEH-reh

You look amazing.

Sei affascinante.
SEH-ee ahf-fah-shee-NAHN-teh

You look like the most interesting person in the room.

Mi sembri la persona più interessante in questo posto.
mee SEHM-bree lah pehr-SOH-nah PYOO een-teh-rehs-SAHN-teh een KWEHS-toh POHS-toh

I wanted to meet you.

Mi piacerebbe conoscerti meglio.
mee pyah-cheh-REHB-beh koh-NOH-shehr-tee MEHL-lyoh

Are you single?	**Sei single?** *SEH-ee SEEN-gohl*
Would you like to dance?	**Ti va di ballare?** *tee vah dee bahl-LAH-reh*
Do you like to dance fast or slow?	**Ti piace il ballo veloce o lento?** *tee PYAH-cheh eel BAHL-loh* *veh-LOH-cheh oh LEHN-toh*
Here, give me your hand.	**Vieni, dammi la mano.** *VYE-nee, DAHM-mee lah MAH-noh*
What would you like to drink?	**Cosa ti va di bere?** *KOH-zah tee vah dee BEH-reh*
You're a great dancer.	**Come balli bene.** *KOH-meh BAHL-lee BEH-neh*
Do you like this song?	**Ti piace questa canzone?** *tee PYAH-cheh KWEHS-tah* *kahn-TSOH-neh*
You have nice eyes!	**Che begli occhi che hai!** *kee BEHL-lyee OHK-kee keh eye*
For body features, see p118.	
May I have your phone number / email address?	**Posso avere il tuo numero di telefono / indirizzo e-mail??** *POHS-soh ah-VEH-reh eel TOO-oh* *NOO-meh-roh dee teh-LEH-foh-noh* *een-dee-REE-tzoh e-mail*
Would you like to go out (with me)?	**Ti va di uscire con me?** *tee vah dee uh-SHEE-reh kohn meh*
I'd love to.	**Mi piacerebbe molto.** *mee pya-cheh-REHB-beh MOHL-toh*

GETTING CLOSER

You're very attractive.

Sei molto bello -a.
say MOHL-toh BEHL-loh -ah

I like being with you.

Mi piace stare con te.
mee PYAH-cheh STAH-reh kohn teh

I like you.

Mi piaci.
mee PYAH-chee

I want to hold you.

Voglio tenerti fra le braccia.
VOHL-lyoh teh-NEHR-tee frah leh BRAHT-chah

Kiss me.

Baciami.
BAH-chah-mee

May I give you a hug / a kiss?

Posso abbracciarti / baciarti?
POHS-soh ahb-braht-CHAHR-tee / bah-CHAHR-tee

Would you like a back rub?

Vuoi che ti massaggi la schiena?
VWOH-ee keh tee mahs-SAHD-jee lah SKYEH-nah

Would you like a massage?

Ti piacerebbe un massaggio?
tee pyah-CHEH-rehb-beh oon mahs-SAHD-joh

NIGHTLIFE

Don't mix the message

Ti desidero / Ti voglio.
tee deh-ZEE-deh-roh /
tee VOHL-lyoh

I desire you / I want you. These are pretty much physical, erotic expressions, much as in English.

Ti amo.
tee AH-moh

This means "I love you" and is used seriously.

GETTING INTIMATE

Would you like to come inside?

Vuoi entrare?
VWOH-ee ehn-TRAH-reh

May I come inside?

Posso entrare?
POHS-soh ehn-TRAH-reh

Let me help you out of that.

Ti aiuto a toglierlo.
tee ah-YOO-toh ah TOHL-lyehr-loh

Would you help me out of this?

Mi aiuti a toglierlo?
mee ah-YOO-tee ah TOHL-lyehr-loh

You smell so good.

Hai un buon profumo.
eye oon bwon proh-FOO-moh

You're beautiful / handsome.

Sei bellissima / bellissimo.
say behl-LEES-see-mah /
beh-LEES-see-moh

May I?

Posso?
POHS-soh

OK?

Va bene?
vah BEH-neh

Like this?

Così?
koh-ZEE

How?

Come?
KOH-meh

HOLD ON A SECOND

Please don't do that.

No, per favore, non farlo.
noh pehr fah-VOH-reh nohn FAHR-loh

Stop, please.

Smetti, per favore.
ZMEHT-tee pehr fah-VOH-reh

Do you want me to stop?

Vuoi che smetta?
VWOH-ee keh ZMEHT-tah

Let's just be friends.

Restiamo solo amici.
reh-STYAH-moh SOH-loh ah-MEE-chee

Do you have a condom?

Hai un preservativo?
eye oon preh-sehr-vah-TEE-voh

Are you on birth control?

Prendi la pillola?
PREHN-dee lah PEEL-loh-lah

I have a condom.

Ho un preservativo.
OH oon preh-sehr-vah-TEE-voh

Do you have anything you should tell me first?

Hai qualcosa da dirmi prima di continuare?
eye kwahl-KOH-zah dah DEER-mee PREE-mah dee kohn-tee-NWAH-reh

BACK TO IT

That's it.	**Ecco, sì.** *EHK-koh SEE*
That's not it.	**No, non così.** *noh nohn koh-ZEE*
Here.	**Qui.** *kwee*
There.	**Lì.** *LEE*
More.	**Ancora.** *ahn-KOH-rah*
Harder.	**Più forte.** *PYOO FOHR-teh*
Faster.	**Più veloce.** *PYOO veh-LOH-cheh*
Deeper	**Più profondo.** *PYOO proh-FOHN-doh*
Slower.	**Più lentamente.** *PYOO lehn-tah-MEHN-teh*
Easy / slowly.	**Piano.** *PYAH-noh*
Enough.	**Basta.** *BAHS-tah*

For a full list of features, see p118.
For a full list of body parts, see p190.

COOLDOWN

You're great.

Sei fantastico -a.
say fahn-TAHS-tee-koh -ah

That was great.

È stato bellissimo.
EH STAH-toh behl-LEES-see-moh

Would you like _____

Vuoi _____
VWOH-ee

  a drink?

  qualcosa da bere?
  kwahl-KOH-zah dah BEH-reh

  a snack?

  qualcosa da mangiare?
  kwahl-KOH-zah dah
  mahn-JAH-reh

  a shower?

  fare la doccia?
  FAH-reh lah DOHT-chah

May I stay here?

Posso stare qui?
POHS-soh STAH-reh kwee

Would you like to stay here?

Vuoi stare qui?
VWOH-ee STAH-reh kwee

I'm sorry. I have to go now.

Mi dispiace. Ora devo andare.
mee dee-SPYAH-cheh OH-rah
DEH-voh ahn-DAH-reh

Where are you going?

Dove vai?
DOH-veh VAH-ee

I have to work early.

Devo alzarmi presto per andare al lavoro.
DEH-voh ahl-TSAHR-mee
PREHS-toh pehr ahn-DAH-reh ahl
lah-VOH-roh

I'm flying home in the morning.

Torno a casa domani mattina.
TOHR-noh ah KAH-zah
doh-MAH-nee maht-TEE-nah

I have an early flight.

Il mio volo parte presto.
eel MEE-oh VOH-loh PAHR-teh
PREHS-toh

NIGHTLIFE

I think this was a mistake.	**Credo che questo sia stato un errore.** *KREH-doh keh KWEHS-toh SEE-ah STAH-toh oon ehr-ROH-reh*
Will you make me breakfast?	**Puoi preparare la colazione per me?** *pwoy preh-pah-RAH-reh lah koh-lah-TSYOH-neh pehr meh*
Stay, I'll make you breakfast.	**Rimani qui, ti preparo la colazione.** *ree-MAH-nee kwee tee preh-PAH-roh lah koh-lah-TSYOH-neh*

IN THE CASINO

How much is this table?	**Quanto costa questo tavolo?** *KWAHN-toh KOHS-tah KWEHS-toh TAH-voh-loh*
Deal me in.	**Entro in gioco.** *EHN-troh een JOH-koh*
Put it on red!	**Sul rosso!** *sool ROHS-soh*
Put it on black!	**Sul nero!** *sool NEH-roh*
Let it ride!	**Lascialo girare!** *LAH-shah-loh jee-RAH-reh*
21!	**Ventuno!** *vehn-TOO-noh*
Snake-eyes!	**Due uno!** *DOO-eh OO-noh*
Seven.	**Sette.** *SEHT-teh*
For a full list of numbers, see p7. Damn, eleven.	**Accidenti, undici.** *at-chee-DEHN-tee OON-dee-chee*
I'll pass.	**Passo.** *PAHS-soh*
Hit me!	**Carte!** *KAHR-te*

Watch that stress!

The meaning of **casinò** (*kah-zee-NOH*) is different from that of **casino** (*kah-ZEE-noh*)! **Casinò** means the gambling house, while **casino** means a mess, a screwed up situation, or . . . "a lot", as in **Mi dispiace un casino** (I'm really sorry about something or someone).

Split.	**Metà e metà.**
	meh-TAH eh meh-TAH
Are the drinks complimentary?	**Le bevande sono gratis?**
	leh beh-VAHN-deh SOH-noh gratis
May I bill it to my room?	**Posso addebitarlo alla mia stanza?**
	POHS-soh ahd-deh-bee-TAHR-loh
	AHL-lah MEE-ah STAHN-tsah
I'd like to cash out.	**Vorrei incassare la vincita.**
	vohr-RAY een-kahs-SAH-reh lah
	VEEN-chee-tah
I'll hold.	**Va bene.**
	vah BEH-neh
I'll see your bet.	**Vedo.**
	VEH-doh
I call.	**Chiamo.**
	KYAH-moh
Full house!	**Full!**
	full
Royal flush.	**Scala reale.**
	SKAH-lah reh-AH-leh
Straight.	**Scala.**
	SKAH-lah

CHAPTER ELEVEN

HEALTH & SAFETY

This chapter covers the terms you'll need to maintain your health and safety—including the most useful phrases for the pharmacy, the doctor's office, and the police station.

AT THE PHARMACY

Please fill this prescription.	**Mi servono questi farmaci, per favore.** *mee SEHR-voh-noh KWEHS-tee FAHR-mah-chee pehr fah-VOH-reh*
Do you have something for ____	**Ha qualcosa per ____** *AH kwahl-KOH-zah pehr*
a cold?	**il raffreddore?** *eel rahf-frehd-DOH-reh*
a cough?	**la tosse?** *lah TOHS-seh*
I need something for ____	**Mi serve qualcosa per ____** *mee SEHR-veh kwahl-KOH-zah pehr*
corns.	**i calli.** *ee KAHL-lee*
congestion.	**la congestione.** *lah kohn-jehs-TYOH-neh*
constipation.	**la stitichezza.** *stee-tee-KEH-tzah*
diarrhea.	**la diarrea.** *lah dyahr-REH-ah*
indigestion.	**l'indigestione.** *leen-dee-jehs-TYOH-neh*
nausea.	**la nausea.** *lah NOW-zeh-ah*
motion sickness.	**il mal d'auto.** *eel mahl DOW-toh*

seasickness.	**il mal di mare.**
	eel mahl dee MAH-reh
acne.	**l'acne.**
	LAHK-neh
warts.	**le verruche.**
	leh vehr-ROO-keh
I need something _____	**Mi serve qualcosa _____**
	mee SEHR-veh kwahl-KOH-zah
to help me sleep.	**aiutarmi a dormire.**
	ah-yoo-TAHR-mee ah
	dohr-MEE-reh
to help me relax.	**aiutarmi a rilassarmi.**
	ah-yoo-TAHR-mee ah
	ree-lahs-SAHR-mee
I want to buy _____	**Vorrei _____**
	vohr-RAY
condoms.	**dei preservativi.**
	day preh-sehr-vah-TEE-vee
an antihistamine.	**un antistaminico.**
	oon ahn-tees-tah-MEE-nee-koh
antibiotic cream.	**una crema antibiotica.**
	OO-nah KREH-mah
	ahn-tee-BYOH-tee-kah
aspirin.	**dell'aspirina.**
	dehl-lahs-pee-REE-nah
non-aspirin pain reliever.	**un analgesico senza aspirina.**
	oon ah-nahl-JEH-zee-koh
	SEHN-tsah ahs-pee-REE-nah
medicine with codeine.	**un farmaco con codeina.**
	oon FAHR-mah-koh kohn
	koh-deh-EE-nah
insect repellant.	**un insettifugo.**
	oohn een-seht-TEE-fuh-goh

HEALTH & SAFETY

AT THE DOCTOR'S OFFICE

I would like to see _____ **Vorrei vedere _____**
vohr-RAY veh-DEH-reh

a doctor. **un medico.**
oon MEH-dee-koh

a chiropractor. **un chiropratico.**
oon kee-roh-PRAH-tee-koh

a gynecologist. **un ginecologo.**
oon jee-neh-KOH-loh-goh

an eye / ears / nose / throat specialist. **un otorinolaringoiatra.**
oon oh-toh-REE-noh-lah-REEN-goh-YAH-trah

a dentist. **un dentista.**
oon dehn-TEES-tah

an optometrist. **un optometrista.**
oon ohp-toh-meh-TREES-tah

Do I need an appointment? **Mi serve un appuntamento?**
mee SEHR-veh oon ahp-poon-tah-MEHN-toh

Do I have to pay upfront? **Devo pagare subito?**
DEH-voh pah-GAH-reh SOO-bee-toh

I have an emergency. **È un'emergenza.**
EH oon-eh-mehr-JEHN-tsah

I need an emergency prescription refill. **Mi servono questi farmaci urgentemente.**
mee SEHR-voh-noh KWEHS-tee FAHR-mah-chee oor-jehn-teh-MEHN-teh

Please call a doctor. **Chiami un medico, per favore.**
KYAH-mee oon MEH-dee-koh pehr fah-VOH-reh

I need an ambulance. **Mi serve un'ambulanza.**
mee SEHR-veh oon-ahm-boo-LAHN-tsah

In an Emergency

In case of a medical emergency, travelers should dial 118 right away.

SYMPTOMS

For a full list of body parts, see p190.

My _____ hurts.

Mi fa male _____.
mee fah MAH-leh

My _____ is stiff.

_____ è rigido -a.
EH REE-jee-doh -ah

I think I'm having a heart attack.

Credo sia un infarto.
KREH-doh SEE-ah oon een-FAHR-toh

I can't move.

Non riesco a muovermi.
nohn ree-EHS-koh ah MWOH-vehr-mee

I fell.

Sono caduto -a.
SOH-noh kah-DOO-toh -ah

I fainted.

Sono svenuto -a.
SOH-noh zveh-NOO-tohlah

I have a cut on my _____.

Ho un taglio su _____.
OH oon TAHL-lyoh soo

I have a headache.

Ho mal di testa.
OH mahl dee TEHS-tah

My vision is blurry

La vista è annebbiata.
lah VEES-tah EH ahn-nehb-BYAH-tah

I feel dizzy.

Mi gira la testa.
mee JEE-rah lah TEHS-tah

I think I'm pregnant.

Credo di essere incinta.
KREH-doh dee EHS-seh-reh een-CHEEN-tah

I don't think I'm pregnant.

Non credo di essere incinta.
nohn KREH-doh dee EHS-seh-reh een-CHEEN-tah

I'm having trouble walking.

Faccio fatica a camminare.
FAHT-choh fah-TEE-kah ah kahm-mee-NAH-reh

I can't get up.

Non riesco ad alzarmi.
nohn ree-EHS-koh ahd ahl-TSAHR-mee

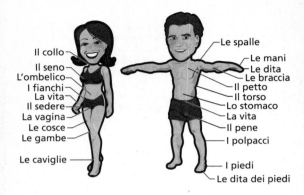

Il collo
Il seno
L'ombelico
I fianchi
La vita
Il sedere
La vagina
Le cosce
Le gambe
Le caviglie

Le spalle
Le mani
Le dita
Le braccia
Il petto
Il torso
Lo stomaco
La vita
Il pene
I polpacci
I piedi
Le dita dei piedi

See p118 for facial features.

I was mugged.	**Sono stato aggredito -a.** *SOH-noh STAH-toh* *ahg-ghreh-DEE-toh -ah*
I was raped.	**Sono stato violentato -a.** *SOH-noh STAH-toh vyo-lehn-* *TAH-toh -tah*
A dog attacked me.	**Un cane mi ha aggredito -a.** *oon KAH-neh mee AH* *ahg-ghreh-DEE-toh -ah*
A snake bit me.	**Mi ha morso un serpente.** *mee AH MOHR-soh oon* *sehr-PEHN-teh*
I can't move my _____ without pain.	**Mi fa male quando muovo _____.** *mee fah MAH-leh KWAHN-doh* *MWOH-voh*
I think I sprained my ankle.	**Credo di essermi slogato la caviglia.** *KREH-doh dee EHS-sehr-mee* *zloh-GAH-toh lah kah-VEEL-lyah*

MEDICATIONS

I need morning-after pills.

Mi serve la pillola del giorno dopo.
*mee SEHR-veh lah PEEL-loh-lah
dehl JOHR-noh DOH-poh*

I need birth control pills.

Mi serve la pillola anticoncezionale.
*mee SEHR-veh lah PEEL-loh-lah
anti-kohn-CHEH--tsyo-NAH-lee*

I need erectile dysfunction
pills.

**Mi servono pillole per la
disfunzione erettile.**
*mee SEHR-voh-noh PEEL-loh-leh
pehr lah dees-foon-TSYOH-neh
eh-REHT-tee-leh*

I lost my eyeglasses and
need new ones.

**Ho perso gli occhiali da vista e me
ne servono di nuovi.**
*OH PEHR-soh lyee ohk-kee-AH-lee
dah VEES-tah eh meh neh
SEHR-voh-noh dee NWOH-vee*

I need new contact lenses.

**Ho bisogno di un nuovo paio di
lenti a contatto.**
*oh bee-SOHN-nyo dee oon NWOH-
voh PAH-yoh dee LEHN-tee ah
kohn-TAHT-toh*

It's cold in here!

Fa freddo qui!
fah FREHD-doh kwee

I am allergic to _____

Sono allergico -a _____
SOH-noh ahl-LEHR-jee-koh -ah

penicillin.

alla penicillina.
AHL-lah peh-nee-cheel-LEE-nah

antibiotics.

agli antibiotici.
AHL-lyee ahn-tee-BYOH-tee-chee

sulfa drugs.

ai sulfonamidi.
eye sool-foh-NAH-mee-dee

steroids.

agli steroidi.
AHL-lee steh-ROY-dee

I have asthma.

Soffro d'asma.
SOHF-froh DAHZ-mah

HEALTH & SAFETY

DENTAL PROBLEMS

Where can I find a dentist?
Dove posso trovare un dentista?
DOH-veh POHS-soh troh-VAH-reh oon dehn-TEES-tah

I have a toothache.
Mi fa male un dente.
mee fah MAH-leh oon DEHN-teh

I chipped a tooth.
Mi si è rotto un dente.
mee see EH ROHT-toh oon DEHN-teh

My bridge came loose.
Mi si è allentato il ponte.
mee see EH ahl-lehn-TAH-toh eel POHN-teh

I lost a crown.
Ho perso una capsula.
OH PEHR-soh OO-nah KAHP-soo-lah

I lost a denture plate.
Ho perso una piastra della dentiera.
OH PEHR-soh OO-nah PYAHS-trah DEHL-lah dehn-TYEH-rah

AT THE POLICE STATION

I'm sorry, did I do something wrong?
Scusi, ho fatto qualcosa di male?
SKOO-zee OH FAHT-toh kwahl-KOH-zah dee MAH-leh

I am ____
Sono ____
SOH-noh

an American.
americano -a.
ah-meh-ree-KAH-noh -ah

a Canadian.
canadese.
kah-nah-DEH-zeh

a European.
europeo -a.
eh-oo-roh-PEH-oh -ah

an Australian.
australiano -a.
ow-strah-LYAH-noh -ah

a New Zealander.
neozelandese.
neh-oh-dzeh-lahn-DEH-zeh

For a full listing of nationalities, see English/Italian dictionary.

Listen Up: Police Lingo

La patente, il libretto e l'assicurazione, per favore. *fah-voh-REES-kah lah pah-TEHN-teh eel lee-BREHT-toh eh LAHS-see-koo-raht-SYOH-neh pehr fah-VOH-reh*	Your license, registration and insurance, please.
La multa è di cinquanta euro. Può pagarla direttamente a me. *lah MOOL-tah EH dee cheen-KWAN-tah EH-oo-roh PWOH pah-GAH-reh dee-reht-tah-MEHN-teh ah meh*	The fine is $50. You can pay me directly.
Il passaporto, per favore. *eel pahs-sah-POHR-toh pehr fah-VOH-reh*	Your passport, please.
Dov'è diretto? *dohv-EH dee-REHT-toh*	Where are you going?
Perchè tanta fretta? *pehr-KEH TAHN-tah FREHT-tah*	Why are you in such a hurry?

The car is a rental.	**L'auto è a noleggio.** *LOW-toh EH ah noh-LEHD-joh*
Do I pay the fine to you?	**Devo pagare la multa a lei?** *DEH-voh pah-GAH-reh lah MOOL-tah ah lay*
Do I have to go to court?	**Devo andare in tribunale?** *DEH-voh ahn-DAH-reh een tree-boo-NAH-leh*
When?	**Quando?** *KWAHN-doh*
I'm sorry, my Italian isn't very good.	**Scusi, non parlo bene l'italiano.** *SKOO-zee nohn PAHR-loh BEH-neh lee-tah-LYAH-noh*

HEALTH & SAFETY

I need an interpreter.	**Mi serve un interprete.** *mee SEHR-veh* *oon-een-TEHR-preh-teh*
I'm sorry, I don't understand the ticket.	**Non capisco il motivo della multa.** *nohn kah-PEES-koh eel moh-TEE-voh DEHL-lah MOOL-tah*
May I call my embassy?	**Posso chiamare la mia ambasciata?** *POHS-soh kyah-MAH-reh lah MEE-ah ahm-bah-SHAH-tah*
I was robbed.	**Mi hanno derubato.** *mee AHN-noh deh-roo-BAH-toh*
I was mugged.	**Sono stato -a aggredito -a.** *SOH-noh STAH-toh -ah ahg-ghreh-DEE-toh -ah*
I was raped.	**Sono stato violentato -a.** *SOH-noh STAH-toh vyo-lehn-TAH-toh -tah*
May I make a report?	**Posso sporgere denuncia?** *POHS-soh SPOHR-jeh-reh deh-NOON-chah*
Somebody broke into my room.	**Qualcuno è entrato nella mia stanza.** *kwahl-KOO-noh EH ehn-TRAH-toh NEHL-lah MEE-ah STAHN-tsah*
Someone stole my ___.	**Mi hanno rubato ___.** *mee AHN-noh roo-BAH-toh*
telephone	**il telefono** *eel teh-LEH-foh-noh*
passport	**il passaporto** *eel PAHS-sah-pohr-toh*
purse	**borsellino** *BOHR-sehl-LEE-noh*
wallet	**portafoglio** *POHR-tah-FOH-lyoh*
computer	**il computer** *eel kohm-PYUH-tehr*
backpack	**lo zaino** *loh DZAY-noh*
camera	**la macchina fotografica** *lah MAHK-kee-nah foh-toh-GRAH-fee-kah*

ENGLISH—ITALIAN

DICTIONARY KEY

n	noun	m	masculine
v	verb	f	feminine
adj	adjective	s	singular
prep	preposition	pl	plural
adv	adverb	pron	pronoun
interj	interjection		

All verbs are listed in infinitive (to + verb) form, cross-referenced to the appropriate conjugations page. Adjectives are listed first in masculine singular form, followed by the feminine ending.

For food terms, see the Menu Reader (p91) and Grocery Section (p100) in Chapter 4, Dining.

A

able, to be able to (can) *v* potere, riuscire a (fare qualcosa), essere in grado di (fare qualcosa) **p30**

above *prep* sopra **p78**

accept, to accept *v* accettare **p20**

Do you accept credit cards? Accettate la carta di credito? **p38**

accident *n* l'incidente *m*

I've had an accident. Ho avuto un incidente.

account *n* il conto *m* **p135**

I'd like to transfer to / from my checking account. Vorrei trasferire dei fondi al / dal mio conto corrente.

I'd like to transfer to / from my savings account. Vorrei trasferire dei fondi al / dal mio conto di risparmio.

acne *n* l'acne *f* **p187**

across *prep* attraverso, dall'altro lato di **p5**

across the street dall'altro lato della strada

actual *adj* reale

adapter plug *n* lo spinotto dell'adattatore *m* **p157**

address *n* l'indirizzo *m* **p123**

What's the address? Qual è l'indirizzo?

admission fee *n* il prezzo d'ingresso *m* **p148**

in advance in anticipo

African-American *adj* afroamericano -a

after *prep* dopo

afternoon n il pomeriggio m
  in the afternoon nel / di
    pomeriggio
age n l'età f **p116**
  What's your age? Quanti
    anni ha?
agency n l'agenzia f
  (travel) l'agenzia viaggi m
agnostic adj agnostico -a
air conditioning n l'aria
    condizionata f **p68**
  **Would you lower / raise
    the air conditioning?** Può
    abbassare / alzare l'aria
    condizionata?
airport n l'aeroporto m
  I need a ride to the airport.
    Ho bisogno di un
    passaggio all'aeroporto.
  **How far is it from the
    airport?** Quanto dista
    dall'aeroporto?
airsickness bag n il sacchetto
    per il mal d'aria m **p48**
aisle (in store) n la corsia f
  Which aisle is it in? In quale
    corsia si trova?
alarm clock n la sveglia f
alcohol n l'alcol m **p88**
  Do you serve alcohol?
    Servite bevande alcoliche?
  I'd like nonalcoholic beer.
    Vorrei una birra analcolica.

all n tutto m **p11**
  all of the time sempre
  That's all, thank you. È
    tutto, grazie.
all adj tutto -a **p11**
allergic adj allergico -a See
    common allergens, **p191**
  I'm allergic to ____. Sono
    allergico -a a ____.
altitude n l'altitudine f
aluminum n l'alluminio m
ambulance n l'ambulanza f
American adj americano -a
amount n la quantità f
angry adj arrabbiato -a
animal n l'animale m
another adj altro -a
answer n la risposta f
answer, to answer v
    rispondere **p20**
  Answer me, please. Per
    cortesia, mi risponda.
antibiotic n l'antibiotico m
  I need an antibiotic. Ho
    bisogno di un antibiotico.
antihistamine n
    l'antistaminico m **p187**
anxious adj ansioso -a
any adj qualsiasi
anything n qualsiasi cosa f
anywhere adv dovunque
appointment n
    l'appuntamento m **p148**

Do I need an appointment?
Serve un appuntamento?

April n aprile m p14

are See **be, to be,** p24

arm(s) n il braccio (s) / le braccia (pl) m

arrival(s) n l'arrivo m / gli arrivi m pl p39

arrive, to arrive v arrivare p20

art n l'arte f

> **exhibit of art** la mostra d'arte
> **art museum** il museo d'arte
> **fine arts** le belle arti
> **Renaissance art** l'arte del Rinascimento

artist n l'artista m f

Asian adj asiatico -a

ask, to ask v chiedere, domandare p20

> **to ask for (to request)** chiedere
> **to ask a question** fare una domanda p28

aspirin n l'aspirina f p187

assist, to assist v assistere p21

assistance n l'assistenza f

asthma n l'asma f p191

I have asthma. Ho l'asma.

at prep a, in

atheist adj, n ateo -a

ATM / cash machine n il bancomat m p135

I'm looking for an ATM / cash machine. Sto cercando un bancomat.

attend, to attend v participare p20

audio adj, n l'audio m p65

August n agosto m p15

aunt n la zia f p115

Australia n l'Australia f

Australian adj australiano -a

autumn n l'autunno m

available adj disponibile

ENGLISH–ITALIAN

B

baby n il / la bambino -a m f p116

baby adj per bambini

> **Do you sell baby food?** Vendete cibo per bambini?

babysitter n il / la baby-sitter m f

> **Do you have babysitters?** Avete baby-sitter?
> **Do you have babysitters who speak English?** Avete delle baby-sitter che parlano inglese?

back n la schiena f p190

> **My back hurts.** Mi fa male la schiena.

back rub n il massaggio alla schiena m p179

backed up (toilet) *adj*
intasato *m*

The toilet is backed up! Il
gabinetto è intasato!

bag *n* la borsa *f*, il sacchetto *m*

airsickness bag il sacchetto
per il mal d'aria.

My bag was stolen. La mia
borsa è stata rubata.

I lost my bag. Ho perso la
mia borsa.

bag, to bag *v* mettere in
borsa p20

baggage *adj, n* il bagaglio *m*

baggage claim il ritiro
bagagli

bait *n* l'esca *f* p168

balance (on bank account) *n*
il saldo *m* p135

balance, to balance *v*
bilanciare p20

balcony *n* il balcone *m*

ball (sport) *n* la palla *f*

ballroom dancing *n* il ballo
da sala *m*

band (musical ensemble) *n* il
gruppo musicale *m*

band-aid *n* il cerotto *m*

bank *n* la banca *f* p133

**Do you know where I can
find a bank?** Sa dov'è una
banca?

bar *n* il bar *m*

barber *n* il barbiere *m*

bass (instrument) *n* il basso *m*

bath *n* il bagno *m*,

bathroom (restroom) *n* il
bagno *m* p62

**Where is the nearest public
bathroom?** Sa dov'è il
bagno pubblico più vicino?

bathtub *n* la vasca da bagno *f*

bathe, to bathe *v* fare il
bagno p28

battery (for flashlight) *n* la
pila *f*

battery (for car) *n* la batteria *f*

bee *n* l'ape *f*

I was stung by a bee. Mi ha
punto un'ape.

be, to be *v* essere p26, stare
p25

beach *n* la spiaggia *f* p170

beach, to beach *v* tirare a
riva p20

beautiful *adj* bello -a p116

bed *n* il letto *m* p67

beer *n* la birra *f* p88

beer on tap la birra alla
spina

begin, to begin *v* cominciare,
iniziare p20

behave, to behave *v*
comportarsi p20, 35

behind *prep, adv* dietro a p5

below *prep, adv* sotto a

belt *n* la cintura *f* p152
  conveyor belt il nastro
  trasportatore
berth *n* la cuccetta *f*
best *adj* il / la migliore
bet *n* la scommessa *f* p184
  I'll see your bet. Eguaglio la
  sua scommessa.
bet, to bet *v* scommettere **p20**
better *adj* migliore
between *prep* fra, tra
big *adj* grande p11
bilingual *adj* bilingue
bill (currency) *n* la banconota *f*
  (check) *n* il conto *m*
  (utility bill) *n* la bolletta *f*
bill, to bill *v* mandare il
  conto **p20**
biography *n* la biografia *f*
biracial *adj* meticcio
bird *n* l'uccello *m*
birth control *n* la
  contraccezione *f* p191
birth control (contraceptive)
  adj anticoncezionale p191
  I'm out of birth control pills.
  Non ho più pillole
  anticoncezionali.
  **I need more birth control
  pills.** Ho bisogno di più
  pillole anticoncezionali.
bit (small amount) *n* un po', un
  pezzetto *m*

black *adj* nero -a
blanket *n* la coperta *f* p47
bleach *n* la candeggina *f*
blind *adj* ipovedente p65
block, to block *v* bloccare **p20**
blond(e) *adj, n* il / la biondo -a
blouse *n* la camicetta *f* p152
blue *adj* azzurro -a, blu
blurry *adj* annebbiato -a
board *n* l'asse *f*
  on board a bordo
board, to board *v* salire a
  bordo di **p21**
boarding pass *n* la carta
  d'imbarco *f* p46
boat *n* l'imbarcazione *f*
bomb *n* la bomba *f*
book *n* il libro *m* p156
bookstore *n* la libreria *f* p156
boss *n* il capo *m*
bottle *n* la bottiglia *f*
  **May I heat this (baby)
  bottle someplace?** Posso
  riscaldare il biberon da
  qualche parte?
box (seat) *n* il tribune *m* p164
box office *n* la biglietteria *f*
boy *n* il ragazzo *m*
boyfriend (friend, date) *n* il
  ragazzo *m*
braille, American *n* il braille
  americano *m*

brake *n* il freno *m* p54

emergency brake il freno
d'emergenza

brake, to brake *v* frenare **p20**

brandy *n* il brandy *m* p89

bread *n* il pane *m*

break, to break *v* rompere **p20**

breakfast *n* la colazione *f*

What time is breakfast? A
che ora è la colazione?

**bridge (across a river, dental
structure)** *n* il ponte *m*

I need a new bridge. Ho
bisogno di un'altra protesi
dentaria.

briefcase *n* la valigetta
portadocumenti *f* p49

bright *adj* brillante, luminoso -a

broadband *n* la banda larga *f*

bronze *adj* bronzo

brother *n* il fratello *m* p115

brown *adj* marrone

brunette *n* la bruna *f*

Buddhist *adj* buddista

budget *n* il bilancio *m*

buffet *n* il buffet *m* p81

bug *n* l'insetto *m*

burn, to burn *v* bruciare,
incendiare **p20**

Can I burn a CD here? Posso
masterizzare un CD qui?

bus *n* l'autobus *m* p60

Where is the bus stop?
Dov'è la fermata
dell'autobus?

Which bus goes to _____?
Quale autobus va a _____?

business *n* l'attività *m* p122

business *adj* commerciale

business center il centro
affari

busy (restaurant) *adj*
affollato -a **(phone)** *adj*
occupato -a

butter *n* il burro *m*

buy, to buy *v* acquistare,
comprare **p20**

C

café *n* il caffè *m*, il bar *m*

Internet café l'Internet café

call, to call (shout) *v* gridare
(telephone) *v* chiamare **p20**

camp, to camp *v*
campeggiare, fare
campeggio **p20, 28**

camper (person) *n* il
campeggiatore / la
campeggiatrice

**Do we need a camping
permit?** Abbiamo bisogno
d'un permesso di
campeggio?

campsite *n* l'area di
campeggio *f* p79

can *n* la scatola *f,* la lattina *f*

can (to be able to) *v* potere p30

Canada *n* il Canada *m*

Canadian *adj, n* canadese *m f*

cancel, to cancel *v* cancellare p20

My flight was canceled. Il mio volo è stato cancellato.

canvas (art) *n* la tela *f*

car *n* l'auto *f*, la macchina *f* See car types, p50.

car rental agency l'autonoleggio

I need a rental car. Mi serve un'auto a noleggio.

card *n* la carta *f* p123

Do you accept credit cards? Accettate le carte di credito?

May I have your business card? Posso avere il suo biglietto da visita?

I'd like a greeting card. Vorrei un biglietto d'auguri.

car seat (child's safety seat) *n* il seggiolino *m*

Do you rent car seats for children? Noleggiate seggiolini per bambini?

car sickness *n* il mal d'auto *m*

cash *n* i contanti *m*, i soldi *m*

cash only solo contanti

cash, to cash *v* incassare p20

to cash out (gambling) incassare la vincita

cash machine / ATM *n* il bancomat *m* p135

cashmere *adj, n* il cashmere *m*

casino *n* il casinò *m* p184

cat *n* il / la gatto -a *m f*

Catholic *adj, n* cattolico -a

cavity (tooth) *n* la carie *f*

I think I have a cavity. Credo di avere una carie.

CD *n* il CD *m* p139

CD player *n* il lettore di CD *m*

celebrate, to celebrate *v* festeggiare p20

cell / mobile phone *n* il cellulare *m* p135

centimeter *n* il centimetro *m*

chamber music *n* la musica da camera *f*

change (money) *n* il resto *m*, gli spiccioli *m pl*

I'd like change, please. Vorrei degli spiccioli, per cortesia.

This isn't the correct change. Il resto non è esatto.

change (to change money or clothes) *v* cambiare p20

changing room *n* il camerino *m*

charge, to charge (money) v
addebitare **(a battery)** v
caricare p20

charmed (greeting) piacere

charred (meat) adj
bruciacchiato -a

charter, to charter v
noleggiare p20

cheap adj economico -a

check n l'assegno m p132

**Do you accept travelers'
checks?** Accettate
i travellers' checks?

check, to check v verificare p20

checked (pattern) adj a
quadretti

check-in (airport) n il
check-in m p38

What time is check-in? A
che ora è il check-in?

check-out (hotel) n check-out
f p78

check-out time l'orario di
check-out (m)

What time is check-out? A
che ora è il check-out?

**check out, to check out
(hotel)** v pagare il conto
dell'albergo p20

cheese n il formaggio m

chicken n il pollo m

child n il / la bambino -a m f

children n i / le bambini -e m f

Are children allowed? Sono
ammessi i bambini?

**Do you have children's
programs?** Avete dei
programmi per bambini?

**Do you have a children's
menu?** Avete un menu per
bambini?

Chinese adj cinese m f

chiropractor n il chiropratico m

chrysanthemum n il
crisantemo m

church n la chiesa f p126

cigarette n la sigaretta f

a pack of cigarettes un
pacchetto di sigarette m

cinema n il cinema m p144

city n la città f p68

claim n il reclamo m

I'd like to file a claim. Vorrei
fare un reclamo.

clarinet n il clarinetto m

class n la classe f p41

business class la Business
class

economy class l'economy
class

first class la prima classe

classical (music, taste) adj
classico -a

clean adj pulito -a

clean, to clean v pulire p21

Please clean the room today. Per favore, oggi pulisca la camera.

clear adj chiaro -a

climb, ascent n la scalata f

climb, to climb v scalare, salire p20, 21

(a mountain) scalare una montagna p20

(stairs) salire le scale

close, to close v chiudere p20

close (near) adj vicino -a

closed adj chiuso -a

cloudy adj nuvoloso -a

clover n il trifoglio m

go clubbing, to go clubbing v andare per locali notturni p27

coat n il cappotto m

coffee n il caffè m p90

iced coffee il caffè freddo

cognac n il cognac m p89

coin n la moneta f p132

cold adj freddo -a p111

cold n il freddo m

I'm cold. Ho freddo.

It's cold. Fa freddo.

cold (infection) n il raffreddore m p186

I have a cold. Ho il raffreddore.

Coliseum n il Colosseo m

collect adj a carico del destinatario

I'd like to place a collect call. Vorrei fare una telefonata a carico ricevente.

collect, to collect v raccogliere p20

college n l' università f

color n il colore m

color, to color v colorare p20

common adj comune

computer n il computer m

concert n il concerto m p130

condition n la condizione f

in good / bad condition in buone / cattive condizioni

condom n il preservativo m p180

Do you have a condom? Hai un preservativo?

Not without a condom. Non senza un preservativo.

confirm, to confirm v confermare p20

I'd like to confirm my reservation. Vorrei confermare la mia prenotazione.

confused adj confuso -a

congested adj congestionato -a

connection speed n la velocità di connessione f

constipated adj stitico -a

I'm constipated. Sono stitico -a.

contact lenses *n* le lenti a contatto *f pl* p191

I lost my contact lenses. Ho perso le lenti a contatto.

continue, to continue *v* continuare p20

convertible *n* le cabriolet *f*

cook, to cook *v* cucinare p20

I'd like a room where I can cook. Vorrei una stanza ad uso cucina.

cookie *n* il biscotto *m*

copper *adj* rame

corner *n* l'angolo *m*

on the corner all'angolo

correct, to correct *v* correggere p20

correct *adj* giusto -a

Am I on the correct train? Mi trovo sul treno giusto?

cost, to cost *v* costare p20

How much does it cost? Quanto costa?

costume *n* il costume *m*

cotton *n* il cotone *m*

cough *n* la tosse *f* p186

cough, to cough *v* tossire p21

counter (board) *n* il bancone *m*

court (legal) *n* il tribunale *m*

court (sport) *n* il campo *m*

courteous *adj* cortese p78

cousin *n* il / la cugino -a *m f*

cover charge (bar, restaurant) *n* il coperto *m* p175

cow *n* la mucca *f*

crack (glass) *n* l'incrinatura *f*

craftsperson *n* l'artigiano -a *m f* p121

cream *n* la crema *f*

credit card *n* la carta di credito *f*

Do you accept credit cards? Accettate carte di credito?

crib *n* la culla *f* p69

crown (dental) *n* la capsula *f*

curb *n* il bordo del marciapiede *m*

curl *n* il ricciolo *m*

curly *adj* riccio -a

currency exchange *n* l'ufficio di cambio *m* p133

Where is the nearest currency exchange? Sa dove si trova un ufficio di cambio più vicino?

current (water, electricity) *n* la corrente *f*

customs *n* la dogana *f* p36

cut (wound) *n* la ferita *f*

I have a bad cut. Ho una brutta ferita.

cut, to cut *v* tagliare p20

cybercafé *n* l'Internet café *m*

Where can I find a cybercafé? Sa dove posso trovare un Internet café?

D

damaged *adj* danneggiato -a

Damn! *expletive* Dannazione!

dance, to dance *v* ballare **p20**

danger *n* il pericolo *m*

dark *adj* scuro -a

dark *n* il buio *m*

daughter *n* la figlia *f* **p114**

day *n* il giorno *m* **p161**

the day before yesterday l'altro ieri

these last few days questi ultimi giorni

dawn *n* l'alba *f*

at dawn all'alba

deaf *adj* ipoudente **p65**

deal (bargain) *n* l'affare *m*

What a great deal! Che affarone!

deal, to deal (cards) *v* dare le carte **p28**

Deal me in. Dia le carte anche a me.

December *n* dicembre *m*

declined *adj* rifiutato -a

Was my credit card declined? La mia carta di credito è stata rifiutata?

declare, to declare *v* dichiarare **p20**

I have nothing to declare. Non ho niente da dichiarare.

deep *adj* profondo -a

delay *n* il ritardo *m* **p44**

How long is the delay? Di quanto è in ritardo?

delighted *adj* felicissimo -a

democracy *n* la democrazia *f*

dent, to dent *v* ammaccare **p20**

He / She dented the car. Lui / Lei ha ammaccato la macchina.

dentist *n* il dentista *m f* **p192**

denture *n* la dentiera *f*

denture plate piastra delle dentiera

departure *n* la partenza *f*

designer *n* lo / la stilista *m f*

dessert *n* il dolce *m* **p98**

dessert menu la lista dei dolci

destination *n* la destinazione *f*

diabetic *adj* diabetico -a **p85**

dial, to dial (phone number) *v* fare il numero **p28**

dial direct fare il numero diretto

diaper *n* il pannolino *m*

Where can I change a diaper? Dove posso cambiare il pannolino?

diarrhea *n* la diarrea *f* **p186**

dictionary *n* il dizionario *m*

different (other) *adj* diverso -a, altro -a

difficult *adj* difficile

dinner *n* la cena *f* p81

directory assistance (phone) *n* l'assistenza telefonica *f*

disability *n* l'invalidità *f*

disappear, to disappear *v* sparire p21

disco *n* la discoteca *f* p174

disconnect, to disconnect *v* staccare p20

disconnected *adj* staccato -a

Operator, I was disconnected. Centralino, è caduta la linea.

discount *n* lo sconto *m*

Do I qualify for a discount? Posso ricevere uno sconto?

dish *n* il piatto *m*

dive, to dive *v* tuffarsi p35

scuba dive fare immersioni subacquee

divorced *adj* divorziato -a

dizzy *adj* stordito -a p189

do, to do *v* fare p28

doctor *n* il medico *m f* p121

doctor's office *n* l'ambulatorio medico *m* p188

dog *n* il cane *m*

service / guide dog il cane guida p65

dollar *n* il dollaro *m* p133

door *n* la porta *f*

double *adj* doppio -a

double bed il letto a due piazze

double vision vederci doppio

down *adv* giù p5

download, to download *v* scaricare p20

downtown *n* il centro città *m*

dozen *n* la dozzina *f*

drain *n* lo scarico *m*

drama *n* il dramma *m*

drawing (art) *n* il disegno *m*

dress (garment) *n* il vestito *m*

dress (general attire) *n* l'abbigliamento *m* p152

What's the dress code? Come ci si deve vestire?

dress, to dress *v* vestirsi p35

Should I dress up for that affair? Mi devo vestire bene per quella occasione?

dressing (salad) *n* il condimento *m*

dried *adj* secco -a

drink *n* la bevanda *f* p88

I'd like a drink. Vorrei qualcosa da bere.

drink, to drink *v* bere p29

drip, to drip *v* sgocciolare p20

drive, to drive *v* guidare p20

driver *n* l'autista *m f*

drum *n* il tamburo *m*

dry *adj* secco -a, asciutto -a

This towel isn't dry. Questo asciugamano non è asciutto.

dry, to dry v asciugarsi **p20**
I need to dry my clothes.
Devo far asciugare i vestiti.
dry cleaner n la lavanderia a
secco f **p74**
dry cleaning n il lavaggio a
secco m
duck n l'anatra f
duty-free adj esente da tasse,
duty-free
duty-free shop n il duty-free
m **p37**
DVD n il DVD m **p157**
**Do the rooms have DVD
players?** C'è il lettore di
DVD nelle camere?
Where can I rent DVDs? Sa
dove posso noleggiare DVD?

E

early adv presto
It's early. E' presto.
eat, to eat v mangiare **p20**
to eat out v mangiare fuori
economy n l'economia f
editor n il redattore m, la
redattrice f **p121**
educator n l'educatore m,
l'educatrice f **p121**
eight adj otto **p7**
eighteen adj diciotto **p7**
eighth adj ottavo -a **p9**
eighty adj ottanta **p7**
election n l'elezione f **p125**

electrical hookup n il
collegamento elettrico m
p79
elevator n l'ascensore m
eleven adj undici **p7**
e-mail n l'e-mail f **p139**
**May I have your e-mail
address?** Posso avere il suo
indirizzo e-mail?
e-mail message il messaggio
e-mail
e-mail, to send e-mail v
inviare un'e-mail **p20**
embarrassed adj
imbarazzato -a
embassy n l'ambasciata f
emergency n l'emergenza f
emergency brake n il freno
d'emergenza m **p54**
emergency exit n l'uscita
d'emergenza f **p41**
employee n il / la
dipendente m f
employer n il datore di lavoro /
la datrice di lavoro m f
engine n il motore m **p53**
engineer n l'ingegnere m f
England n l'Inghilterra f
English n, adj inglese m f
Do you speak English? Parla
inglese? **p2**
enjoy, to enjoy v piacere **p33**
enter, to enter v entrare **p20**
Do not enter. Ingresso
vietato.

enthusiastic *adj* entusiasta

entrance *n* l'entrata *f*

envelope *n* la busta *f*

environment *n* l'ambiente *m*

escalator *n* la scala mobile *f*

espresso *n* il caffè *m*

evening *n* la sera *f*

exchange rate *n* il cambio *m*

What is the exchange rate
for U.S. / Canadian dollars?
Qual è il cambio del
dollaro USA / canadese?

excuse, to excuse (pardon) *v*
scusare **p20**

Excuse me. Mi scusi.
(to get through) Permesso.

exhausted *adj* esausto -a

exhibit *n* la mostra *f*

exit *n* l'uscita *f* **p39**

not an exit senza uscita

exit, to exit *v* uscire **p32**

expensive *adj* caro -a **p175**

explain, to explain *v* spiegare
p20

express *adj* espresso -a

express check-in il check-in
espresso **p40**

extra (additional) *adj* extra /
in più

extra-large *adj* extra-large

eye *n* l'occhio *m* **p190**

eyebrow *n* il sopracciglio *m*

eyeglasses *n* gli occhiali *m*

eyelashes *n* le ciglia *f pl*

F

fabric *n* il tessuto *m*

face *n* il viso *m* **p118**

faint, to faint *v* svenire **p21**

fall (season) *n* l'autunno *m*

fall, to fall *v* cadere **p20**

family *n* la famiglia *f* **p114**

fan *n* il ventilatore *m*

far *adj* lontano -a **p5**

How far is it to _____?
Quanto dista _____?

fare *n* la tariffa *f* **p52**

fast *adj* veloce

fat *adj* grasso -a **p11**

fat *n* il grasso *m* **p11**

father *n* il padre *m* **p115**

faucet *n* il rubinetto *m*

fault *n* la colpa *f*

I'm at fault. E' colpa mia.
It was his fault. E' colpa sua.

fax *n* il fax *m* **p123**

February *n* febbraio *m* **p14**

fee *n* l'onorario *m,* la tassa *f*

female *adj* femminile *f*

female *n* la donna /
la femmina *f*

fiancé(e) *n* il / la
fidanzato -a *m f* **p115**

fifteen *adj* quindici **p7**

fifth *adj* quinto -a **p9**

fifty *adj* cinquanta **p7**

find, to find *v* trovare **p20**

fine (traffic violation) *n* la
multa *f*

fine *adj* bello -a p1
  I'm fine (well). Sto bene.
fire *n* il fuoco *m*
  Fire! Al fuoco!
first *adj* primo -a p9
fishing pole *n* la canna da
  pesca *f* p168
fitness center *n* il centro
  benessere *m* p66, 161
fit, to fit (size) *v* andare bene
  p27 **(looks)** *v* stare bene
  p25, 35
  This doesn't fit. Questo non
  mi va bene.
  Does this look like it fits?
  Mi sta bene?
fitting room *n* il camerino *m*
five *adj* cinque p7
flight *n* il volo p38
  **Where do domestic flights
  arrive?** Dove arrivano i voli
  nazionali?
  **Where do domestic flights
  depart?** Da dove partono i
  voli nazionali?
  **Where do international
  flights arrive?** Dove arrivano
  i voli internazionali?
  **Where do international
  flights depart?** Da dove
  partono i voli internazionali?

  **What time does this flight
  leave?** A che ora parte
  questo volo?
flight attendant l'assistente
  di volo *m f*
floor *n* il piano *m* **(ground)** *n*
  il pavimento *m*
  ground floor il pianoterra
  first floor il primo piano
flower *n* il fiore *m*
flush (gambling) *n* il flush *m*
flush, to flush *v* tirare l'acqua
  del water p20
  This toilet won't flush. Non
  funziona lo sciacquone.
flute *n* il flauto *m*
food *n* il cibo *m* p91
foot (body part) *n* il piede *m*
for *prep* per
forehead *n* la fronte *f* p118
format *n* il formato *m*
formula *n* la formula *f*
  **Do you sell infants'
  formula?** Vendete il latte
  in polvere?
forty *adj* quaranta p7
forward *adv* avanti p6
four *adj* quattro p7
fourteen *adj* quattordici p7
fourth *adj* quarto -a p9
  one-fourth un quarto *m*
fragile *adj* fragile

freckle n la lentiggine f

French adj francese m f

fresh adj fresco -a p109

Friday n venerdì m p14

friend n l'amico -a, m f

from prep da

front adj anteriore adv davanti p42

front desk la reception

front door la porta principale

fruit n il frutto m p104 (collective) n la frutta f

fruit juice n il succo di frutta m See fruits, p104.

full adj pieno -a

Full house! n Full house! f

fuse n il fusibile m

G

garlic n l'aglio m

gas n il gas m p53 (fuel) n la benzina f

gas gauge la spia del serbatoio p54

out of gas la benzina è finita

gate (at airport) n il gate m

German adj, n tedesco -a m f

gift n il regalo m

girl n la ragazza f

girlfriend n la ragazza f p114

give, to give v dare p28

glass (drinking) n il bicchiere m

Do you have it by the glass? Lo servite a bicchiere?

I'd like a glass please. Vorrei un bicchiere per favore.

glass (material) n il vetro m

glasses (spectacles) n gli occhiali m, pl p191

I need new glasses. Ho bisogno di un nuovo paio di occhiali.

glove n il guanto m

go, to go v andare p27

goal (sport) n il goal m

goalie n il portiere m

gold adj oro

golf n il golf m p172

golf, to go golfing v giocare a golf p20

good adj buono -a

goodbye n arrivederci m See common salutations, p111.

goose n l'oca f

grade (school) n la classe f

gram n il grammo m

grandfather n il nonno m

grandmother n la nonna f

grandparents n i nonni m pl

grape n l'uva f

gray *adj* grigio -a
Great! *adj* Eccellente!
Greek *adj* greco -a
Greek Orthodox *adj* greco-ortodosso -a
green *adj* verde
groceries *n* la spesa *f* p100
group *n* il gruppo *m*
grow, to grow (get larger) *v* crescere **p20**
  Where did you grow up? Dov'è cresciuto -a?
guard *n* la guardia *f* p37
  security guard la guardia di sicurezza
guest *n* l'ospite *m f*
guide (tour) *n* la guida *f*
(publication) *n* la guida *f*
guide, to guide *v* guidare **p20**
guided tour *n* la visita guidata *f* p148
guitar *n* la chitarra *m*
gym *n* la palestra *f* p161
gynecologist *n* il / la ginecologo -a *m f*

H

hair *n* i capelli *m pl* p118
haircut *n* il taglio di capelli *m*
  I need a haircut. Ho bisogno di tagliare i capelli.
  How much is a haircut? Quanto costa il taglio?

hairdresser *n* il / la parrucchiere -a *m f* p158
hair dryer *n* l'asciugacapelli *m*
half *adj* mezzo -a
half *n* la metà *f*
hallway *n* il corridoio *m*
hand *n* la mano *f*, le mani *f, pl*
handbag *n* la borsetta *f*
handicapped-accessible *adj* accessibile ai disabili
handle, to handle *v* maneggiare **p20**
  Handle with care. Maneggiare con cura.
handsome *adj* bello -a p116
hangout (hot spot) *n* il ritrovo *m*
hang out, to hang out (relax) *v* rilassarsi **p20, 35**
hang up, to hang up (end a phone call) *v* riattaccare **p20**
hanger *n* la gruccia *f*
happy *adj* felice p120
hard *adj* duro -a
hat *n* il cappello *m*, il berretto *m*
have, to have *v* avere **p27**
hazel *adj* color nocciola
hazelnut *n* la nocciola *f*
headache *n* il mal di testa *m*

headlight n il faro della macchina m

headphones n le cuffie f pl

hear, to hear v udire, sentire **p21**

hearing-impaired adj ipoudente p65

heart n il cuore m

heart attack n l'infarto m

hectare n l'ettaro m 10

Hello! n Salve! / Ciao! See greetings, p111.

Help! n Aiuto!

help, to help v aiutare **p20**

hen n la gallina f

her pron lei f p19

her, hers adj, pron suo -a, suoi / sue pl p19

herb n l'erbetta f

here adv qui, qua p5

high adj alto -a

highlights (hair) n i colpi di sole m p158

highway n l'autostrada f

hike, to hike v fare un'escursione a piedi **p28**

him pron lui m p19

Hindu adj indù

hip-hop n l'hip-hop m p174

his adj, pron suo -a, sing, suoi / sue pl p19

historical adj storico -a

history n la storia f

hobby n l'hobby m

hold, to hold v reggere **p20**

to hold hands tenersi per mano

Would you hold this for me? Può reggermi questo, per favore?

hold, to hold (wait) v aspettare **p20**

Hold on a minute! Aspetti un attimo!

I'll hold. Sì, attendo.

holiday n la festa f, la vacanza f

to go on holiday andare in vacanza

home n la casa f

homemaker n il casalingo m / la casalinga f

horn n il corno m

horse n il cavallo m

hostel n l'ostello m p66

hot adj caldo -a

hot chocolate n la cioccolata calda f p89

hotel n l'albergo, l'hotel m

Do you have a list of local hotels? Ha un elenco di alberghi locali?

hour n l'ora f p12

hours (schedule) n l'orario m

how adv come p3

humid adj umido -a p124

hundred adj cento p7

hurry, to hurry v aver
   fretta **p27**
   I'm in a hurry. Ho fretta.
   Hurry, please! Si sbrighi per
   favore!
hurt, to hurt v far male **p28**
   Ouch! That hurts! Ahi! Fa
   male!
husband n il marito m **p114**

I

I pron io **p19**
ice n il ghiaccio m
identification n il documento
   di riconoscimento m
in prep in
indigestion n l'indigestione f
inexpensive adj economico -a
infant n il / la neonato -a m f
   Are infants allowed? Si
   possono portare i neonati?
information n l'informazione f
information booth n il punto
   informazioni m
injury n la ferita f **p188**
insect repellent n
   l'insettifugo, m **p187**
inside adj interno -a
inside adv dentro
insult, to insult v insultare **p20**
insurance n l'assicurazione f
intercourse (sexual) n il
   rapporto sessuale m **p180**

interest rate n il tasso
   d'interesse m
intermission n l'intervallo m
Internet n l'Internet m **p139**
   High-speed Internet
   Connessione Internet
   veloce
   **Do you have Internet
   access?** Avete l'accesso a
   Internet?
   **Where can I find an Internet
   café?** Sa dove posso
   trovare un Internet café?
interpreter n l'interprete m f
   I need an interpreter. Mi
   serve un interprete.
introduce, to introduce v
   presentare **p20**
   **I'd like to introduce you to
   ____.** Ho il piacere di
   presentarle ____.
Ireland n l'Irlanda f
Irish adj irlandese m f
is See be, to be, **p24.**
Italian adj italiano -a

J

jacket n la giacca f **p152**
January n gennaio m **p14**
Japanese adj giapponese
jazz n il jazz m
Jewish adj ebreo -a
jog, to run v correre **p20**

juice *n* il succo *m*

July *n* luglio *m* p15

June *n* giugno *m* p14

K

keep, to keep *v* tenere, conservare p20

kid *n* il / la ragazzo -a *m f*

Are kids allowed? Sono ammessi i bambini?

Do you have kids' programs? Avete programmi per bambini?

Do you have a kids' menu? Avete un menu per i bambini?

kilo *n* il chilo *m*

kilometer *n* il chilometro *m*

kind (type) *n* il tipo *m*

(nice) *adj* gentile, simpatico -a

What kind is it? Che tipo è?

kiss *n* il bacio *m* p180

kitchen *n* la cucina *f* p73

know, to know (something) *v* sapere p31

know, to know (someone) *v* conoscere p20

kosher *adj* kasher p85

L

lactose-intolerant *adj* intollerante al lattosio

land, to land *v* atterrare p20

landscape (painting) *n* il paesaggio *m*, **(land)** *n* il panorama *m*

language *n* la lingua *f*

laptop *n* il portatile *m* p139

large *adj* grande p11

last, to last *v* durare p20

last *adj* ultimo -a

late *adv* tardi p13

Please don't be late. Non fare tardi per favore.

later *adv* più tardi p4

See you later. A più tardi.

lately *adv* recentemente

laundry (shop) *n* la lavanderia *f*, **(clothes)** il bucato *m*

lavender *adj* color lavanda

law *n* la legge *f*

lawyer *n* l'avvocato *m* / l'avvocatessa *f* p121

least *n* il minimo *m*, *adj* minimo -a

leather *n* la pelle *f*

leave, to leave (depart) *v* partire p21

left *adj* sinistro -a p5

on the left a sinistra

(remaining) rimasto -a

leg *n* la gamba *f* p190

lemonade *n* la limonata *f*

less *adv* meno p10

lesson *n* la lezione *f*

license *n* il permesso *m* p50
   driver's license la patente *f*
life *n* la vita *f*
   the good life la dolce vita *f*
life preserver *n* il salvagente *m*
light (brightness) *adj*
   luminoso -a
light *n* la luce *f* p47
   (for cigarette) l'accendino
   May I offer you a light?
   Posso offrirle da accendere?
   (lamp) la lampada *f*
   (weight) leggero -a
like, to like *v* piacere p32
   I would like ____. Mi piacerebbe / Vorrei ____.
   I like this place. Mi piace
   questo posto.
limo *n* la limousine *f*
liqueur *n* il liquore *m* p89
liquor *n* il liquore *m* p89
liter *n* il litro *m* p10
little *adj* piccolo -a
live, to live *v* vivere p20
   (dwell) *v* abitare p20
   Where do you live? Dove
   abita?
   What do you do for a living?
   Che lavoro fa?
local *adj* locale
lock *n* la serratura *f* p54
lock, to lock *v* chiudere a
   chiave p20

I can't lock the door. Non
   riesco a chiudere la porta a
   chiave.
   I'm locked out. Sono
   rimasto chiuso fuori.
locker *n* l'armadietto *m* p162
   storage locker l'armadietto
   di deposito
   locker room lo spogliatoio
long (length) *adj* lungo -a p10
   adv lungo, molto tempo
   For how long? Per quanto
   tempo?
   long ago molto tempo fa
look, to look (to observe) *v*
   guardare, osservare p20
   I'm just looking. Sto solo
   guardando.
   Look here! Guarda qui!
   look (to appear) *v* sembrare
   p20
   How does this look? Come
   ti sembra questo?
look for, to look for (to search)
   v cercare p20
   I'm looking for a porter. Sto
   cercando un facchino.
loose *adj* sciolto -a
lose, to lose *v* perdere p20
   I lost my passport. Ho perso
   il passaporto.
   I lost my wallet. Ho perso il
   portafogli.
   I'm lost. Mi sono perso.

ENGLISH–ITALIAN

lost *adj* perso -a p46
loud *adj* rumoroso -a p78
loudly (voice) *adv* ad alta voce
lounge *n* la sala d'aspetto *f*
lounge, to lounge *v* bighellonare p20
love *n* l'amore *m*
love, to love *v* amare p20
   (family) voler bene
   (a friend) voler bene
   (a lover) amare
   to make love fare l'amore p28
low *adj* basso -a p5
lunch *n* il pranzo *m*
luggage *n* il bagaglio *m* p48
   Where do I report lost luggage? Dove posso denunciare la perdita del bagaglio?
   Where is the lost luggage claim? Dov'è l'ufficio bagagli smarriti?

M
machine *m* la macchina *f*
made of *adj* fatto -a di
magazine *n* la rivista *f*
maid (hotel) *n* la cameriera *f*
maiden *adj* nubile *f*
   That's my maiden name. E' il mio nome da nubile.

mail *n* la posta *f* p141
   air mail la posta aerea
   registered mail posta assicurata
make, to make *v* fare p28
makeup *n* il trucco *m*
make up, to make up (apply cosmetics) *v* truccarsi p20, 35
make up, to make up (apologize) *v* rimediare p28
male *adj* maschile *m*
male *n* il maschio *m*
mall *n* il centro commerciale *m*
man *n* l'uomo *m*
manager *n* il manager *m f*, il direttore / il direttrice *m f*
manual *n* il manuale *m*
many *adj* molti -e
map *n* la cartina *f* p55
March *n* marzo *m* p15
market *n* il mercato *m* p154
   flea market il mercatino
   open-air market il mercato all'aperto
married *adj* sposato -a p116
marry, to marry *v* sposarsi p20, 35
massage, to massage *v* massaggiare p20
match (sport) *n* la partita *f*
   (stick) il fiammifero
   book of matches una bustina di fiammiferi

match, to match *v* abbinare **p20**
> **Does this ____ match my outfit?** Questo -a ___ si abbina al mio completo? **p152**

May *n* maggio *m*
may *v* potere **p30**
> **May I ____?** Posso ____?

meal *n* il pasto *m*
meat *n* la carne *f* **p95**
meatball *n* la polpetta *f*
medication *n* il farmaco *m*
medium (size) *adj* medio -a
medium rare (meat) *adj* quasi al sangue
medium well (meat) *adj* ben cotto -a
member *n* il socio *m*
menu *n* il menu *m* **p91**
> **May I see a menu?** Posso vedere il menu?
> **children's menu** il menu dei bambini
> **diabetic menu** il menu per i diabetici
> **kosher menu** il menu kasher
> **vegetarian menu** il menu vegetariano

metal detector *n* il metal detector *m*
meter *n* il metro *m*
middle *adj* medio -a **p10**
midnight *n* la mezzanotte *f*

mile *n* il miglio *m*
military *n* il militare *m*
milk *n* il latte *m* **p88**
> **milk shake** *n* il frappè *m*

milliliter *n* il millilitro *m* **p10**
millimeter *n* il millimetro *m*
minute *n* il minuto *m* **p12**
> **in a minute** fra un minuto

miss, to miss (a flight) *v* perdere **p20**
missing *adj* perso -a
mistake *n* l'errore *m*
moderately priced *adj* a prezzo modico *m* **p66**
mole (facial feature) *n* il neo *m*
Monday *n* lunedì *m* **p13**
money *n* il denaro *m*, i soldi *m pl* **p132**
> **money transfer** il bonifico bancario **p132**

month *n* il mese *m* **p15**
morning *n* il mattino *m* **p13**
> **in the morning** al mattino

mosque *n* la moschea *f* **p126**
mother *n* la madre *f* **p114**
mother, to mother *v* coccolare **p20**
motorcycle *n* la moto *f*
mountain *n* la montagna *f*
> **mountain climbing** alpinismo

mouse *n* il topo *m*
mouth *n* la bocca *f* **p118**

move (change position) v spostare, spostarsi **p20, 35**

(relocate) v traslocare **p20**

movie n il film m **p144**

much adj molto -a

to get mugged essere vittima di un assalto **p190**

museum n il museo m **p148**

music n la musica f **p128**

live music la musica dal vivo

musician n il musicista m f

Muslim adj musulmano -a

my / mine pron mio -a, miei m pl, mie f pl **p19**

mystery (novel) n il giallo m

N

name n il nome m **p114**

My name is ___. Mi chiamo ___.

What's your name? Come si chiama?

napkin n il tovagliolo m

narrow adj stretto -a

nationality n la nazionalità f

nausea n la nausea f **p186**

near adj vicino **p5**

nearby adv vicino **p5**

neat (tidy) adj ordinato -a

need, to need v avere bisogno di qualcosa **p20**

I need ___. Ho bisogno di ___.

neighbor n il / la vicino -a m f

nephew n il nipote m **p115**

network n la rete f

new adj nuovo -a

newspaper n il giornale m

newsstand n l'edicola f **p38**

New Zealand n la Nuova Zelanda f

New Zealander adj neozelandese m f

next adj prossimo -a

next to prep accanto a **p5**

the next station la prossima stazione

nice adj simpatico -a

niece n la nipote f **p113**

night n la notte f **p13**

at night di notte

per night a notte

nightclub n il locale notturno m **p174**

nine adj nove **p7**

nineteen adj diciannove **p7**

ninety adj novanta **p7**

ninth adj nono -a **p9**

no adv no **p1**

noisy adj rumoroso -a

none pron nessuno **p10**

no smoking adj vietato fumare

nonsmoking area la zona non-fumatori

nonsmoking room la stanza per non-fumatori

noon n il mezzogiorno m

nose n il naso m **p190**

novel *n* il romanzo *m*

not *adv* non

nothing *n* il niente *m*

November *n* novembre *m*

now *adv* adesso p4

number *n* il numero *m* p7

Which room number? Che numero di camera?

May I have your phone number? Posso avere il suo numero di telefono?

nurse *n* l'infermiere -a *m f*

nurse, to nurse (breastfeed) *v* allattare p20

Do you have a place where I can nurse? C'è un posto dove posso allattare?

nursery *n* l'asilo nido *m*

Do you have a nursery? Avete un asilo nido?

nut *n* la noce *f*

O

o'clock *adv* in punto

two o'clock le due in punto

October *n* ottobre *m*

of *prep* di

offer, to offer *v* offrire p21

officer *n* il poliziotto *m*

off-white *adj* bianco, sporco

oil *n* l'olio *m*

okay *adv* okay, va bene p2

old *adj* vecchio -a

olive *n* l'oliva *f*

on *prep* su, sopra

one *adj* uno -a p7

one way (traffic sign) *n* il senso unico *m*

open (business) *adj* aperto -a

Are you open? Siete aperti?

opera *n* l'opera *f* p146

operator (phone) *n* il centralino *m* p136 l'operatore/l'operatrice *m f*

optometrist *n* l'optometrista *m*

orange *adj* arancio

orange juice *n* il succo d'arancia *m* p88

order, to order (demand) *v* ordinare **(request)** *v* chiedere p20

organic *adj* organico -a, **(food)** biologico -a

Ouch! *interj* Ahi!

ours *pron* nostro -a, nostri -e

out *adv* fuori

outside *adj* esterno -a

over *prep* sopra, su

overcooked *adj* troppo cotto -a

overheat, to overheat *v* surriscaldare p20

The car overheated. La macchina si è surriscaldata.

overflowing *adj* traboccante

oxygen tank *n* la bombola d'ossigeno *f* p64

ENGLISH—ITALIAN

P

package n il pacco m p141

pacifier n il ciuccio m

page, to page (someone) v far chiamare v p28

paint, to paint v dipingere p20

painting n il quadro m, la pittura f

pale adj pallido -a

paper n la carta f

parade n la parata f

parent n il genitore / la genitrice m f p115

park n il parco m

park, to park v parcheggiare p20

parking n il parcheggio m

 no parking sosta vietata

 parking fee la tariffa del parcheggio

 parking garage il garage

partner n il / la compagno -a m f

party n il party m, la festa f

 political party il partito politico

pass, to pass v passare p20

 I'll pass. Io passo.

passenger n il / la passeggero -a m f

passport n il passaporto m

 I've lost my passport. Ho perso il passaporto.

password n la password f

past adj passato

past n il passato m

 (in space) prep dopo, oltre

pay, to pay v pagare p20

peanut n l'arachide f, la nocciolina f

pedestrian n il pedone m

pediatrician n il / la pediatra m f

 Can you recommend a pediatrician? Può consigliarmi un pediatra?

permit n il permesso m

 Do we need a permit? Abbiamo bisogno di un permesso?

permit, to permit v permettere p20

petrol / gas n la benzina f

phone n il telefono m p135

 Do you have a phone directory? Ha un elenco telefonico?

 May I have your phone number? Può darmi il suo numero di telefono?

 Where can I find a public phone? Dove posso trovare un telefono pubblico?

 phone operator l'operatore m, l'operatrice f

 Do you sell prepaid phones? Vendete i telefonini con scheda prepagata?

ENGLISH—ITALIAN

phone call *n* la telefonata *f*
I need to make a collect phone call. Ho bisogno di fare una telefonata a carico del ricevente.
an international phone call una telefonata internazionale *f*
photocopy, to photocopy *v* fotocopiare **p20**
piano *n* il pianoforte *m*
pillow *n* il cuscino *m* p47
down pillow il cuscino di piuma *m*
pink *adj* rosa
pizza *n* la pizza *f* p92
place, to place *v* mettere **p20**
plastic *n* la plastica *f*
play *n* il gioco *m*
play, to play (a game) *v* giocare, **(an instrument)** *v* suonare **p20**
playground *n* il parco giochi *m*
Do you have a playground? Avete un parco giochi?
plaza *n* la piazza *f*
please (polite entreaty) *interj* per favore **p1**
please, to be pleasing to *v* accontentare **p20**
pleasure *n* il piacere *m*
It's a pleasure. E' un piacere.
plug (electrical) *n* la spina *f*
plug, to plug *v* inserire la spina **p21**

point, to point *v* indicare **p20**
Would you point me in the direction of ____? Può indicarmi la direzione per ____?
police *n* la polizia *f* p192
police station *n* la stazione di polizia *f*
pool *n* la piscina *f* **(game)** *n* il biliardo *m*
pop music *n* la musica pop *f*
popular *adj* popolare
port (beverage, harbor) *n* il porto *m*
porter *n* il facchino *m* il portiere **(concierge)** *m*
portion *n* la porzione *f*
portrait *n* il ritratto *m*
postcard *n* la cartolina *f*
post office *n* l'ufficio postale *m*
Where is the post office? Dov'è l'ufficio postale?
poultry *n* il pollame *m*
prefer, to prefer *v* preferire **p32** (like uscire)
pregnant *adj* incinta *f*
prepared *adj* preparato -a
prescription *n* la ricetta *f*
price *n* il prezzo *m*
print, to print *v* stampare **p20**
private berth / cabin *n* la cuccetta / cabina privata *f*
problem *n* il problema *m*
process, to process *v* elaborare **p20**

product *n* il prodotto *m*

professional *adj* professionale

program *n* il programma *m*

May I have a program?
Posso avere un
programma?

Protestant *adj* protestante

publisher *n* l'editore *m* /
l'editrice *f*

pull, to pull *v* tirare **p20**

pump *n* la pompa *f*

purple *adj* viola

purse *n* la borsetta *f*

push, to push *v* spingere **p20**

put, to put *v* mettere **p20**

Q

quarter *n* il quarto *m*

one-quarter un quarto *m*

quick *adj* veloce

quiet *adj* tranquillo

R

rabbit *n* il coniglio *m*

radio *n* la radio *f* **p51**

satellite radio la radio
satellitare

rain, to rain *v* piovere **p20**

Is it supposed to rain? E'
prevista pioggia?

rainy *adj* piovoso -a **p124**

It's rainy. Piove.

ramp *n* la rampa *f* **p65**

rare (meat) *adj* al sangue

rate (fee) *n* la tariffa *f* **p52**

What's the rate per day?
Qual è la tariffa
giornaliera?

What's the rate per week?
Qual è la tariffa
settimanale?

rate plan (cell phone) *n* il
piano tariffario *m*

rather *adv* preferibilmente

raven *n* il corvo *m*

read, to read *v* leggere **p20**

really *adv* davvero

receipt *n* la ricevuta *f*, lo
scontrino *m* **p133**

receive, to receive *v* ricevere
p20

recommend, to recommend *v*
raccomandare, consigliare
p20

red *adj* rosso -a

redhead *n* dai capelli rossi *m f*

reef *n* la scogliera *f* **p169**

refill, to refill (beverage) *v*
riempire **p21**

refill (of prescription) la
ripetizione di una ricetta

reggae *n* la musica reggae *f*

relative *n* il / la parente *m f*

remove, to remove *v* togliere
p20

rent, to rent *v* noleggiare **p20**

I'd like to rent a car. Vorrei
noleggiare un'auto.

repeat, to repeat v ripetere p20

Would you please repeat that? Può ripeterlo per favore?

reservation n la prenotazione f p43

I'd like to make a reservation for ___. Vorrei fare una prenotazione per ___. See numbers, p7.

restaurant n il ristorante m

Where can I find a good restaurant? Dove posso trovare un buon ristorante? See restaurant types, p80.

restroom n il bagno m p62

Do you have a public restroom / toilet ? C'è un bagno pubblico?

return, to return (to a place) v ritornare p20

return, to return (something to a store) v restituire p20

ride, to ride v viaggiare p20

right adj il / la destro -a p5

It is on the right. E' sulla destra.

Turn right at the corner. All'angolo, giri a destra.

rights n i diritti m pl

civil rights i diritti civili

river n il fiume m

road n la strada f p55

road closed sign n il segnale della strada bloccata f

rob, to rob v rubare p20

I've been robbed. Sono stato derubato.

rock climbing n la scalata rocciosa f

rocks n le rocce f

I'd like it on the rocks. Lo gradirei con ghiaccio.

romance (novel) n il romanzo m

romantic adj romantico -a

room (hotel) n la camera, la stanza f p72

room for one / two la camera singola / doppia

room service il servizio in camera

rope n la fune f

rose n la rosa f

royal flush n il royal flush m

run, to run v correre p20

S

sad adj triste p120

safe (container) n la cassaforte f p75

Do the rooms have safes? C'è una cassaforte nelle camere?

safe (secure) adj sicuro -a

Is this area safe? Questa zona è sicura?

sail *n* la vela *f* p168

sail, to sail *v* navigare
a vela **p20**

When do we sail? Quando
salpiamo?

salad *n* l'insalata *f* p91

salesperson *n* il / la
commesso -a *m f* p122

salt *n* il sale *m* p85

Is that low-salt? Questo è
con poco sale?

satellite *n* il satellite *m* p68

satellite radio la radio
satellitare

satellite tracking il tracking
satellitare

Saturday *n* il sabato *m* p14

sauce *n* il sugo *m* p93

say, to say *v* dire **p28**

scan, to scan *v* (with a
scanner) scannerizzare **p20**

schedule *n* l'orario *m*

school *n* la scuola *f*

scooter *n* lo scooter *m* p50

score *n* il punteggio *m*

Scottish *adj* scozzese

scratch *n* il graffio *m*

scratch, to scratch *v* graffiare
p20

scratched *adj* graffiato -a

scratched surface superficie
graffiata

scuba dive, to scuba dive *v*
immergersi con le
bombole d'ossigeno **p20,
35**

sculpture *n* la scultura *f*

seafood *n* i frutti di mare *m pl*

search *n* la ricerca *f*

hand search la
perquisizione manuale

search, to search *v* cercare
p20

seasick *adj* che soffre il mal
di mare **p62**

I am seasick. Ho mal di
mare.

seasickness pill *n* la pillola
contro il mal di mare *f*

seat *n* il sedile *m*, il posto *m*

child seat il seggiolino per
bambini

second *adj* secondo -a **p9**

security *n* la sicurezza *f*

security checkpoint il
controllo sicurezza

security guard la guardia di
sicurezza

see, to see *v* vedere **p20**

May I see it? Posso vederlo?

self-serve *n* il self-service *m f*

sell, to sell *v* vendere **p20**

seltzer *n* la soda *f* p88

send, to send *v* spedire,
mandare, inviare **p20, 21**

separated (marital status) adj separato -a p116

September n settembre m

serve, to serve v servire **p21**

service n il servizio m

out of service fuori servizio

services (religious) n le funzioni religiose f

service charge / cover n il coperto m

seven adj sette **p7**

seventy adj settanta **p7**

seventeen adj diciassette **p7**

seventh n adj il / la settimo -a **p9**

sew, to sew v cucire **p21**

sex n il sesso m

sex, to have (intercourse) v avere rapporti sessuali **p27**

shallow adj poco profondo -a

sheet (bed linen) n il lenzuolo m, le lenzuola f

(paper) n il foglio m

shellfish n i crostacei m pl

ship n la nave f **p62**

ship, to ship v spedire **p21**

How much to ship this to ____? Quanto costa per spedire questo a ____?

shipwreck n il naufragio m

shirt n la camicia f **p152**

shoe n la scarpa f **p152**

shop n il negozio m

shop, to shop v fare compere **p28**

I'm shopping for mens' clothes. Sto cercando abbigliamento da uomo.

I'm shopping for womens' clothes. Sto cercando abbigliamento da donna.

I'm shopping for childrens' clothes. Sto cercando abbigliamento per bambini.

short adj basso -a **p10**

shorts n i calzoncini corti m

shot (liquor) n il bicchierino m

shout, to shout v gridare **p20**

show (performance) n lo spettacolo m

What time is the show? A che ora inizia lo spettacolo?

show, to show v mostrare **p20**

Would you show me? Può mostrarmelo?

shower n la doccia f **p71**

Does it have a shower? C'è una doccia?

shower, to shower v fare la doccia **p28**

shrimp n il gambero m

shuttle bus n l'autobus navetta m

sick adj malato -a **p48**

I feel sick. Mi sento male.

side n il lato m

on the side (separately) a parte

sidewalk n il marciapiede m
sightseeing n la gita turistica f
sightseeing bus n l'autobus turistico m
sign, to sign v firmare **p20**
Where do I sign? Dove devo firmare?
silk n la seta f
silver n l'argento m
silver adj argento
sing, to sing v cantare **p20**
single (unmarried) adj single, celibe m, nubile f **p116**
Are you single? E' single?
single (one) adj singolo -a
single bed il letto ad una piazza
sink n il lavabo m
sister n la sorella f **p115**
sit, to sit v sedersi **p20, 35**
six adj sei **p7**
sixteen adj sedici **p7**
sixty adj sessanta **p7**
size (clothing) n la taglia f
(shoes) n il numero m
skin n la pelle f
sleeping berth n la cuccetta f
slow adj lento -a
slow, to slow v rallentare **p20**
Slow down! Rallenti!
slowly adv lentamente
Please speak more slowly. Per favore parli più lentamente.

slum n i bassifondi m pl
small adj piccolo -a **p11**
smell, to smell v puzzare **p20**
smoke, to smoke v fumare **p20**
smoking n il fumo m **p37**
smoking area la zona fumatori
No Smoking Vietato fumare
snack n lo spuntino m
Snake eyes! n Occhi di serpe!
snorkel, to snorkel v fare snorkeling **p28, 169**
soap n il sapone m
sock n il calzino m **p152**
soda n il selz m **p88**
diet soda la bibita dietetica
soft adj morbido -a
software n il software m
sold out adj tutto esaurito
some adj qualche **p10**
someone n qualcuno -a m f
something n qualcosa
son n il figlio m **p114**
song n la canzone f
sorry adj pentito -a
I'm sorry. Mi dispiace.
soup n la zuppa f, la minestra f **p93**
spa n le terme f pl
Spain n la Spagna f
Spanish adj spagnolo -a
spare tire n la ruota di scorta f

speak, to speak v parlare **p20**
  Do you speak English? Parla inglese?
  Would you speak louder, please? Può parlare a voce più alta, per favore?
  Would you speak slower, please? Per favore, può parlare più lentamente?
special (featured meal) n la specialità del giorno f
specify, to specify v specificare **p20**
speed limit n il limite di velocità m **p56**
  What's the speed limit? Qual è il limite di velocità?
speedometer n il tachimetro m **p54**
spell, to spell v scrivere correttamente **p20**
  How do you spell that? Come si scrive?
spice n la spezia f
spill, to spill v versare, rovesciare **p20**
split (gambling) n la divisione f
sport n lo sport m **p163**
spring (season) n la primavera f **p15**
stadium n lo stadio m **p163**
staff (personnel) n il personale m

stamp (postage) n il francobollo m
stair n la scala f
  Where are the stairs? Dove sono le scale?
  Are there many stairs? Ci sono molte scale?
stand, to stand v stare in piedi **p24, 25**
start, to start (commence) v iniziare, cominciare **p20**
start, to start (a car) v mettere in moto **p20**
state n lo stato m
station n la stazione f **p58**
  Where is the nearest gas station? Dov'è il distributore di benzina più vicino?
  Where is the nearest bus station? Dov'è la stazione degli autobus più vicina?
  Where is the nearest subway station? Dov'è la stazione della metropolitana più vicina?
  Where is the nearest train station? Dov'è la stazione ferroviaria più vicina?
stay, to stay v restare **p20**
  We'll be staying for ____ nights. Pernotteremo per ____ notti. *See numbers,* p7.
steakhouse n il ristorante specializzato in bistecche m

steal, to steal v rubare **p20**

stolen adj rubato -a **p49**

stop n la fermata f

Is this my stop? E' questa la mia fermata?

I missed my stop. Ho passato la mia fermata.

stop, to stop v fermare **p20**

Please stop. Per favore si fermi.

Stop, thief! Fermo, al ladro!

store n il negozio m

straight adj diritto -a **p5**

straight ahead avanti diritto

(drink) liscio **p88**

Go straight (directions). Vada diritto.

straight (gambling) n la scala f

street n la via f, la strada f

across the street dall'altra parte della strada

down the street in fondo alla strada

Which street? Quale strada?

How many more streets? Quante strade ancora?

stressed adj stressato -a

striped adj a strisce

stroller n il passeggino m

Do you rent baby strollers? Noleggiate passeggini?

substitution n la sostituzione f

suburb n la periferia f

subway / underground n la metropolitana f, il metrò m **p63**

subway line la linea della metropolitana

subway station la stazione della metropolitana

Which subway do I take for ____? Che linea del metrò devo prendere per ____?

subtitle n il sottotitolo m

suitcase n la valigia f **p49**

suite n la suite m **p66**

summer n l'estate f **p15**

sun n il sole m

sunburn n la scottatura f

I have a bad sunburn. Ho una brutta scottatura.

Sunday n la domenica f **p13**

sunglasses n gli occhiali da sole m pl

sunny adj soleggiato -a **p124**

It's sunny out. C'è il sole fuori.

sunroof n il tettuccio apribile m

sunscreen / sunblock n la crema solare protettiva f

Do you have sunscreen / sunblock SPF ____? Avete la crema solare protettiva fattore ____? See numbers, **p7**.

supermarket *n* il supermarket *m* p100

surf, to surf *v* fare il surf **p28**

surfboard *n* la tavola da surf *f*

suspiciously *adv* in modo sospetto

swallow, to swallow *v* inghiottire **p21**

sweater *n* la maglia *f* p152

swim, to swim *v* nuotare **p20**

Can one swim here? Si può nuotare qui?

swimsuit il costume da bagno

swim trunks il costume da bagno

symphony *n* la sinfonia *f*

T

table *n* il tavolo *m* p82

table for two un tavolo per due

tailor *n* il sarto *m*

Can you recommend a good tailor? Può consigliarmi un buon sarto?

take, to take *v* portare **p20**

Take me to the station. Mi porti alla stazione.

How much to take me to ____? Quanto costa andare a ____?

takeout / takeaway menu *n* il menu da asporto *m*

talk, to talk *v* parlare **p20**

tall *adj* alto -a

tan *adj* marroncino

tanned *adj* abbronzato -a

taste (flavor) *n* il sapore *m*, il gusto *m*

taste (discernment) *n* il gusto *m*

taste, to taste *v* assaggiare **p20**

tax *n* la tassa *f* p155

value-added tax (VAT) l'imposta sul valore aggiunto (IVA)

taxi *n* il taxi / il tassì *m* p57

Taxi! Taxi!

Would you call me a taxi? Mi chiama un taxi per favore?

tea *n* il tè *m* p90

team *n* la squadra *f* p164

techno *n* la musica techno *f*

television *n* la televisione *f*

temple *n* il tempio *m* p126

ten *adj* dieci p7

tennis *n* il tennis *m* p67

tennis court il campo da tennis

tent *n* la tenda *f* p79

tenth *n adj* il / la decimo -a p9

terminal (airport) *n* il terminal *m* p36

Thank you. *interj* Grazie. p1

that *adj* quello -a, quelli -e *pl*

theater *n* il teatro *m* p144

their(s) *pron* di loro; il / la / i / le loro *m f*

them *pron* loro *m f* p19

there (demonstrative) *adv* là

Is / Are there ___? C'è / Ci sono ___?

It's there. E' là.

these *adj* questi -e *m f* p6

thick *adj* spesso -a

thin *adj* sottile

third *adj* terzo -a p9

thirteen *adj* tredici p7

thirty *adj* trenta p7

this *adj* questo -a p6

those *adj* quelli -e p6

thousand *adj* mille p7

three *adj* tre p7

Thursday *n* il giovedì *m* p14

ticket *n* il biglietto *m* p36

ticket counter la biglietteria

one-way ticket un biglietto di sola andata

round-trip ticket un biglietto di andata e ritorno

tight *adj* stretto -a

time *n* il tempo *m*, l'ora *f*

Is it on time? E' in orario?

At what time? A che ora?

What time is it? Che ora è?

timetable (train) *n* l'orario *m*

tip (gratuity) *n* la mancia *f*

tire *n* la gomma *f*, lo pneumatico *m* p54

I have a flat tire. Ho una gomma a terra.

tired *adj* stanco -a p120

to *prep* a; per

today *adv* oggi p4

toilet *n* il gabinetto *m*, la toilette *f*

The toilet is overflowing. Il gabinetto trabocca.

The toilet is backed up. Il gabinetto è intasato.

toilet paper *n* la carta igienica *f*

You're out of toilet paper. E' finita la carta igienica.

toiletries *n* articoli da toeletta *m pl*

toll *n* il pedaggio *m*

tomorrow *adv* domani p4

ton *n* la tonnellata *f*

too (excessively) *adv* troppo

too (also) *adv* anche

tooth *n* il dente *m* p192

I lost my tooth. Ho perso un dente.

toothache *n* il mal di denti *m*

I have a toothache. Ho il mal di denti.

total *n* il totale *m*

What is the total? Qual è il totale?

tour *n* la gita *f* p148

Are guided tours available?
Ci sono visite guidate?

Are audio tours available?
Ci sono visite con audio-guida?

towel *n* l'asciugamano *m*

May we have more towels?
Possiamo avere altri asciugamani?

toy *n* il giocattolo *m*

toy store il negozio di giocattoli

Do you have any toys for the children? Avete dei giocattoli per i bambini?

traffic *n* il traffico *m*

How's traffic? Com'è il traffico?

Traffic is terrible. Il traffico è orribile.

traffic rules le regole del traffico *f*

trail *n* il sentiero *m* p165

Are there trails? Ci sono dei sentieri?

train *n* il treno *m* p58

express train l'espresso

local train il locale

Does the train go to _____?
Questo treno va a _____?

May I have a train schedule? Posso avere un orario del treno?

Where is the train station?
Dov'è la stazione dei treni?

train, to train *v* addestrare p20

transfer, to transfer *v*
trasferire **p32** (like uscire)

I need to transfer funds.
Devo trasferire del denaro.

transmission *n* il cambio *m*

automatic transmission il cambio automatico

standard transmission il cambio manuale

travel, to travel *v* viaggiare p20

trim, to trim (hair) *v* spuntare p20

trip *n* il viaggio *m* p62

triple *adj* triplo -a

trumpet *n* la tromba *f*

trunk / boot (car) *n* il portabagagli *m* p54

try, to try (attempt) *v* cercare di **(clothing)** *v* provare **(food)** *v* assaggiare p20

Tuesday *n* il martedì *m* p15

turkey *n* il tacchino *m*

turn, to turn *v* girare p20

to turn left / right girare a sinistra / destra p20

to turn off / on spegnere / accendere p20

twelve *adj* dodici p7

twenty *adj* venti p7

twine *n* lo spago *m*

two *adj* due p7

U

umbrella *n* l'ombrello *m*

uncle *n* lo zio *m* p115

under *prep* sotto p5

undercooked *adj* crudo -a

understand, to understand *v* capire p32 (like uscire)

I don't understand. Non capisco.

Do you understand? Capisce? / Capite?

underwear *n* la biancheria intima *f*

university *n* l'università *f*

up *adv* su, sopra p5

update, to update *v* aggiornare p20

upgrade *n* la categoria superiore *f* p52

upload, to upload *v* caricare p20

upscale *adj* di lusso

us *pron* noi p3

USB port *n* la porta USB *f*

use, to use *v* usare p20

V

vacation *n* la vacanza *f* p44

on vacation in vacanza

to go on vacation andare in ferie p27

vacancy *n* la disponibilità *f*

van *n* il furgoncino *m* p50

vegetable *n* la verdura *f* p106

vegetarian *adj* vegetariano -a

vending machine *n* il distributore automatico *m*

version *n* la versione *f*

very *adj* molto -a

video *n* il video *m*

Where can I rent videos or DVDs? Dove posso noleggiare videocassette o DVD?

view *n* la vista *f* p69

beach view la vista sulla spiaggia

city view la vista sulla città

vineyard *n* il vigneto *f*

vinyl *n* il vinile *m*

violin *n* il violino *m*

visa *n* il visto *m*

Do I need a visa? Ho bisogno del visto?

vision *n* la visione *f*

visit, to visit *v* visitare p20

visually impaired *n* l'ipovedente *m*

vodka *n* la vodka *f* p88

voucher *n* il buono *m*

W

wait, to wait v attendere **p20**

Please wait. Per favore attenda.

wait n attesa f

How long is the wait? Quanto c'è da aspettare?

waiter n il / la cameriere -a m f

waiting area n la sala d'aspetto m **p36**

wake-up call n il servizio di sveglia f **p75**

wallet n il portafogli m **p46**

I lost my wallet. Ho perso il portafogli.

Someone stole my wallet. Mi hanno rubato il portafogli.

walk, to walk v camminare **p20**

walker (device) n il deambulatore m

walkway n il passaggio pedonale m

moving walkway la passerella mobile

want, to want v volere **p31**

war n la guerra f **p125**

warm adj caldo -a **p124**

watch, to watch v guardare

p20

water n l'acqua f **p46**

Is the water potable? L'acqua è potabile?

Is there running water? C'è l'acqua corrente?

wave, to wave v salutare con la mano **p20**

waxing n la depilazione con ceretta f **p159**

weapon n l'arma f

wear, to wear v indossare **p20**

weather forecast n le previsioni del tempo f pl

Wednesday n il mercoledì m

week n la settimana f **p14**

this week questa settimana

last week la settimana scorsa

next week la settimana prossima

last week la settimana scorsa

weigh, to weigh v pesare **p20**

I weigh ____ kilos. Peso ____ chili.

It weighs ____. Pesa ____. See p7 for numbers.

weights n i pesi m pl

welcome adv benvenuto -a

You're welcome. Prego.

well *adv* bene

  well done (meat) ben cotto -a

  well done (task) ben fatto -a

  I don't feel well. Non mi sento bene.

western *adj* il western

whale *n* la balena *f*

what *adv* che p3

  What sort of ____? Che tipo di ____?

  What time is ____? Che ora è ____? p7

wheelchair *n* la sedia a rotelle *f* p64

  wheelchair access l'accesso per la sedia a rotelle

  wheelchair ramp la rampa di accesso per la sedia a rotelle

  power wheelchair la sedia a rotelle motorizzata

when *adv* quando *See p3 for questions.*

where *adv* dove p3

  Where is it? Dov'è? / Dove si trova? *See p3 for questions.*

which *adv* quale p3

  Which one? Quale? *See p3 for questions.*

white *adj* bianco -a

who *adv* chi p3

whose *adj* di chi

wide *adj* largo -a

widow, widower *n* la / il vedova -o *m f* p116

wife *n* la moglie *f* p114

wi-fi *n* il wi-fi *m* p139

window *n* la finestra *f*

  drop-off window lo sportello di consegna

  pick-up window lo sportello di ritiro

windshield / windscreen *n* il parabrezza *m* p54

windshield wiper *n* il tergicristallo *m* p54

windsurf, to windsurf *v* fare il windsurf p28

windy *adj* ventoso -a p124

wine *n* il vino *m* p88

winter *n* l'inverno *m* p15

wiper *n* il tergicristallo *m*

with *prep* con

withdraw, to withdraw *v* ritirare p20

  I need to withdraw money. Ho bisogno di prelevare dei soldi.

without *prep* senza

woman *n* la donna *f*

work, to work *v* lavorare p20

  This doesn't work. Questo non funziona.

workout n l'esercizio fisico m

worse adj peggiore

worst adj il peggiore

write, to write v scrivere p20

Would you write that down for me? Me lo può scrivere per favore?

writer n lo scrittore m, la scrittrice f p121

X

x-ray machine n l'apparecchiatura per le radiografie f

Y

yellow adj giallo -a

yes interj sì p1

yesterday adv ieri p4

the day before yesterday ieri l'altro

yield sign n il segnale di precedenza m p56

you (sing. informal) tu p19

you (sing. formal) Lei

you (pl informal) voi

you (pl formal) Loro, Voi

your(s) (informal) pron tuo -a m f, tuoi m pl, tue f pl

your(s) (pl informal) pron vostro -a m f, vostri -e m f pl

yours (formal) pron di lei; suo -a m f, suoi m pl, sue f pl

yours (pl formal) pron di loro; il / la loro m f

young adj giovane

Z

zoo n lo zoo m p130

ENGLISH—ITALIAN

A

l'abbigliamento *m clothes n* (general attire) p152

abbinare *v to match* **p20**

abbronzato -a *tanned adj*

abitare *v to live* **p20**

> Dove abita? *Where do you live?*
>
> Abito con il / la mio -a ragazzo -a. *I live with my boyfriend / girlfriend.*

accendere *v to start* (a car), *to turn on* **p20**

> Posso offrirle da accendere? *May I offer you a light?*

l'accendino *m light, lighter n* (cigarette)

l'accesso ai disabili *m access for disabled n*

accettare *v to accept* **p20**

> Si accettano carte di credito. *Credit cards are accepted.*

accettato -a *accepted adj*

l'accettazione *f check-in n*

l'acconciatura afro *f afro n*

accontentare *v to please, to be pleasing to* **p20**

l'acne *f acne n* p187

l'acqua *f water n* p142

> l'acqua calda *hot water*
> l'acqua fredda *cold water*

acquistare *v to buy* **p20**

addestrare *v to train* **p20**

adesso *now adv*

l'aeroporto *m airport n* p36

l'affare *m business, deal, bargain n*

> Fatti gli affari tuoi. *Mind your own business.*

affittare *v to rent* **p20**

> Vorrei affittare _____. *I'd like to rent _____. See p162 for sporting equipment.*

affollato -a *crowded, busy adj* (restaurant)

l'afroamericano -a *African American adj*

l'agenzia *f agency n*

aggiornare *v to update* **p20**

l'aglio *m garlic n*

agnostico -a *m f agnostic n*

agosto *m August n* p14

Ahi! *Ouch! interj*

l'aiuto *m help n*

> Aiuto! *m Help! n*

aiutare *v to help* **p20**

l'alba *f dawn n* p13

> all'alba *at dawn*

albergo *m hotel n* p66

l'alcol *m alcohol n* p88

allattare *v to nurse* (breastfeed) **p20**

l'**allergia** f allergy n

allergico -a allergic adj p191

l'**alluminio** m aluminum n

l'**altitudine** f altitude n

alto -a tall, high, loud adj

più alto -a higher, taller

il / la più alto -a the highest, tallest

altro -a (an)other adj

alzare v to lift, raise, turn up (sound) p20

amare v to love p20

l'**ambasciata** f embassy n

l'**ambiente** m environment n

l'**ambulanza** f ambulance n

l'**ambulatorio del medico** m doctor's office n

l'**americano -a** American adj n

l'**amico -a** m f friend n

ammaccare v to dent p20

l'**ammaccatura** f dent n

l'**amore** m love n

l'**anatra** f duck n

anche too, also adv

andare v to go p27

andare a ballare to go clubbing p174

andare bene to fit

l'**angolo** m corner n

all'angolo on the corner

l'**animale** m animal n

annebbiato -a blurry adj

anni m pl years n (age)

Quanti anni hai? What's your age?

l'**antibiotico** m antibiotic n

l'**anticipo** m advance n

in anticipo in advance

anticoncezionale birth control adj p191

Non ho più pillole anticoncezionali. I'm out of birth control pills.

l'**antistaminico** m antihistamine n p187

l'**ape** f bee n

aperto -a open adj

l'**appartenenza ad un'associazione** f membership n

approvare v approve p20

La sua carta di credito non è stata accettata. Your credit card has been declined.

l'**appuntamento** m appointment n p148

aprile m April n p15

l'**arachide** f peanut n

l'**arancia** m orange n (fruit)

arancione orange adj (color)

ITALIAN–ENGLISH

l'area di campeggio f
    campsite n p79
l'argento m silver n
argento silver adj (color)
l'aria f air n
l'aria condizionata f air
    conditioning n p68
l'arma f weapon n
l'armadietto m locker n p162
  l'armadietto della palestra
    gym locker
  l'armadietto di deposito
    storage locker
arrivare v to arrive p20
Arrivederci. See you later. interj
l'arrivo m, gli arrivi m pl
    arrival n, arrivals n pl p39
l'arte f art n
  le belle arti fine arts
  il museo d'arte art museum
  la mostra d'arte art exhibit
articoli da toeletta m pl
    toiletries n
l'artista m f artist n
l'ascensore m elevator, lift n
l'asciugacapelli m hair dryer n
l'asciugamano m towel n
asciugare v to dry p20
asciutto -a dry, dried adj
l'asiatico -a Asian adj n
l'asilo nido m nursery n
l'asino m donkey n

l'asma f asthma n p191
aspettare v to wait, hold p20
l'aspirina f aspirin n p187
assaggiare v to try (food), to
    taste p20
l'asse f board n
l'assegno m check n p132
l'assicurazione f insurance n
  l'assicurazione sugli incidenti
    collision insurance
  assicurazione sulla
    responsabilità liability
    insurance
l'assistenza f assistance n
l'assistenza telefonica f
    directory assistance n (phone)
assistere v to attend, to assist p20
l'ateo -a atheist adj n
attendere v to wait p20
  Attenda, per favore. Please
    wait.
atterrare v to land p20
l'attesa f wait n
l'attico m penthouse n
attraverso across prep p5
l'audio m audio adj n p65
  Per favore alza l'audio.
    Please turn up the audio.
Auguri! Best wishes! interj
l'auricolare m headphones n
l'australiano -a Australian
    adj n

l'autista *m driver n*

l'auto *f car n* p50

l'autonoleggio *car rental agency* p50

l'autobus *m bus n* p60

la fermata dell'autobus *bus stop*

l'autobus navetta *shuttle bus*

l'autobus turistico *sightseeing bus n*

l'automobile *f car n* p50

l'autunno *m autumn n* p15

l'autostrada *f highway n*

avanti *forward adj* p6

avere *v to have* p27

avere rapporti sessuali *to have sex*

avere fretta *to hurry*

avere bisogno *v to need* p20

Ho bisogno di ____. *I need (to) ____.*

l'avvocato -essa *m f lawyer n* p121

azzurro -a *light blue adj*

B

il / la baby-sitter *m f babysitter n*

il bacio *m kiss n* p180

il bagaglio *m*, i bagagli *m pl luggage, baggage n* p48

i bagagli smarriti *lost baggage*

il recupero bagagli *baggage claim*

il bagno *m bath, bathroom, restroom n* p62

il bagno delle donne *women's restrooms*

il bagno degli uomini *men's restrooms*

il balcone *m balcony n*

ballare *v to dance* p20

il ballo *m dance* p178

il ballo da sala *m ballroom dancing n*

i bambini *pl children n pl*

cibo per bambini *baby food*

passeggini *baby strollers*

il / la bambino -a *m f baby, child n* p116

la banca *f bank n* p133

bancario -a *bank adj*

il conto bancario *bank account*

la tessera bancaria *bank card*

il banco *m counter n* (bar)

il bancomat *m ATM / cash machine n* p135

la banconota *f bill n* (currency) p132

la **banda larga** f *broadband* n
il **bar** m *bar, café* n
il **barattolo** m *tin can* n
il **barbiere** m *barber* n p158
la **barca** f *boat* n
basso -a *short, low* adj p10
  più basso -a *lower*
  il / la più basso -a *lowest*
il **basso** m *bass* n (instrument)
i **bassifondi** m pl *slum* n
bello -a *beautiful, handsome*
  adj p116
bene *fine, well* adv p2
  Sto bene. *I'm fine.*
Benvenuto. *Welcome.* interj
  Benvenuti in Italia.
    Welcome to Italy.
la **benzina** f *gasoline, petrol* n
  indicatore di benzina *gas*
    gauge p54
  La benzina è finita. *It's out*
    of gas.
  Dov'è il distributore di
    benzina più vicino? *Where*
    is the nearest gas station?
bere v *to drink* p29
la **bevanda** f *drink* n p88
  bevanda offerta dalla casa
    complimentary drink
  Desidero una bevanda. *I'd*
    like a drink.

la **bevanda alcolica** f
  alcoholic drink n p88
la **biancheria intima** f
  underwear n
bianco -a *white* adj
  bianco sporco *off-white* adj
il **biberon** m *feeding bottle* n
la **bibita analcolica** f *soda,*
  soft drink n p88
  la bibita dietetica f *diet soda*
la **biblioteca** f *library* n
il **bicchiere** m *glass* n
  Lo servite a bicchiere? *Do*
    you have it by the glass?
  Vorrei un bicchiere per
    favore. *I'd like a glass*
    please.
un bicchierino m *shot* n
  (liquor)
bighellonare v *to lounge* p20
la **biglietteria** f *ticket*
  counter, box office n
il **biglietto** m *ticket* n p36
  un biglietto di sola andata
    one-way ticket
  un biglietto di andata e
    ritorno *round-trip ticket*
  il biglietto da visita *business*
    card
bilanciare v *to balance* p20
bilanciato -a *balanced* adj
il **bilancio** m *budget* n

il **biliardo** m pool n (game)
bilingue bilingual adj
il / la **biondo -a** blond(e) adj n
la **birra** f beer n p88
  birra alla spina draft beer
il **biscotto** m cookie, biscuit n
bloccare v to block p20
il **blocco** m block n
la **bocca** f mouth n p118
la **bomba** f bomb n
la **bombola d'ossigeno** f oxygen tank n p64
il **bordo della strada** m curb n
  a bordo on board
la **borsa** f bag n
la **borsetta** f purse, handbag n
il **botteghino** m box office n
la **bottiglia** f bottle n
il **box** m garage n
il **braccio** m arm n p190
il **Braille** m Braille n
brillante bright adj
il **brillante** m diamond n
bronzo bronze adj
bruciacchiato -a charred (food) adj
bruciare v to burn p20
bruno -a brown adj
  la **bruna** f brunette n
il / la **brutto -a** m f ugly n
il **bucato** m laundry n
il / la **buddista** Buddhist adj n

il **buffet** m buffet n p80
buio -a dark adj
  il **buio** m dark n
buono -a good, fine adj
  buon giorno good morning
  buon pomeriggio good afternoon
  buona sera good evening
  buona notte good night
il **buono** m voucher n
  il **buono per i pasti** meal voucher
  il **buono per la camera** room voucher
il **burro** m butter n
la **busta** f envelope n

C

la **cabina privata** f private berth, cabin n
cadere v to fall p20
il **caffè** m café, coffee n p90
  il **caffè freddo** iced coffee
caldo -a warm, hot adj p124
il **calzino** m sock n p152
i **calzoncini corti** m pl shorts n
cambiare v to change p20
  cambiarsi v to change clothes p35
il **cambio** m change n
  il **cambio della valuta** m exchange rate n
  il **cambio di moneta** m money exchange n

il cambio di velocità *f transmission n*

il cambio di velocità automatico *automatic transmission*

il cambio di velocità manuale *standard transmission*

il cambiavalute *m currency exchange n* p132

la camera d'albergo *f hotel room n* p72

il cameriere *m waiter n*

la cameriera *f waitress, maid n*

il camerino *m changing room n*

la camicetta *f blouse n* p152

la camicia *f shirt n* p152

camminare *v to walk* p20

il cammino *m walk, path n*

campeggiare *v to camp, to go camping* p20

il camper *m camper n*

il campo *m field n* (sport)

canadese *Canadian adj n*

cancellare *v to cancel* p20

la candeggina *f bleach n*

il cane *m dog n*

il cane guida *service dog* p65

la canna da pesca *f fishing pole n* p168

il canovaccio *m dish towel n*

cantare *v to sing* p20

la canzone *f song n*

i capelli *m pl hair n* p118

dai capelli rossi *redhead adj*

capire *v to understand* p32

Non capisco. *I don't understand.*

Capisce / Capite? *Do you understand?*

il capo *m boss n*

il cappello *m hat n*

il cappotto *m coat n*

la capra *f goat n*

il carabiniere *m policeman n*

caricare *v to charge* (a battery), *to load* p20

caricare dati *v to upload* p21

a carico del destinatario *collect adj*

la carie *f tooth cavity n* p192

la carne *f meat n* p95

caro -a *dear, expensive adj*

la carta *f paper, card n* p123

la carta igienica *f toilet paper*

il piatto di carta *paper plate*

il tovagliolo di carta *paper napkin*

Avete un mazzo di carte? *Do you have a deck of cards?*

la carta di credito *credit card*

Si accettano carte di credito. *Credit cards accepted.*

la carta d'imbarco f *boarding pass* p46
la cartina geografica f *map n*
la carta geografica di bordo *onboard map*
la cartolina f *postcard n*
la casa f *home, house n*
a casa mia *at / in my home*
la casa editrice *publishing house*
la casalinga f *homemaker n*
il casinò m *casino n* p184
la cassaforte f *safe n* (container) p75
il / la cattolico -a *Catholic adj n*
il cavallo m *horse n*
il CD m *CD n* p139
la cena f *dinner, supper n*
il centimetro m *centimeter n*
cento m *hundred n* p7
il centralino m *operator n* (phone) p136
il centro m *center n*
in centro m *downtown*
il centro benessere *fitness center* p66, 161
il centro commerciale *mall*
il centro affari m *business center*
cercare v *to look for, to attempt* p20
la ceretta f *waxing n* p150

che *what interj* p3
Che tipo di ____? *What sort of ____?*
chi *who pron* p3
di chi *whose* p3
chiamare v *to call, to shout* p20
chiamare all'altoparlante v *to page* (someone) p20
la chiamata a carico del destinatario *collect phone call*
chiarire v *to clear* p21
chiaro -a *clear adj*
la chiave f *key n*
chiedere v *to order* (request), *to ask for* p20
chiedere scusa v *to apologize* p20
la chiesa f *church n* p126
il chilo m *kilo n*
il chilometro m *kilometer n*
il chiropratico m *chiropractor n*
la chitarra f *guitar n*
chiudere v *to close* p20
chiudere a chiave v *to lock* p20
chiuso -a *closed adj*
il cibo m *food n* p91
il cibo in scatola m *canned goods n*
cieco -a *blind adj* p65

le ciglia *f* eyelashes *n*

il cigno *m* swan *n*

il cinema *m* cinema *n* p144

il / la cinese Chinese *adj n*

cinquanta fifty *adj* p7

cinque five *adj* p7

la cintura *f* belt *n* p152

la cioccolata calda *f* hot chocolate *n* p89

la città *f* city *n* p68

in città downtown

il ciuccio *m* pacifier *n*

il clarinetto *m* clarinet *n*

la classe *f* class *n* p41

business class business class

l'economy class economy class

la prima classe first class

il climatizzatore *m* air conditioning *n*

la colazione *f* breakfast *n*

la prima colazione *f* breakfast *n*

il collegamento elettrico *m* electrical hookup *n* p79

colorare *v* to color p20

il colore *m* color *n*

il Colosseo *m* Coliseum *n*

la colpa *f* fault *n*

i colpi di sole *m* highlights *n* (hair) p158

come how *adv* p3

cominciare *v* to begin p20

commerciale business *adj*

il / la commesso -a *m f* salesperson *n*

il / la compagno -a *m f* partner *n*

comportarsi *v* to behave p35

comprare *v* to buy p20

compreso -a included *adj*

comune common *adj*

con with *prep*

il concerto *m* concert *n* p130

il condimento *m* salad dressing *n*

la condizione *f* condition *n*

in buone / cattive condizioni in good / bad condition

la conferma *f* confirmation *n*

confermare *v* to confirm p20

confuso -a confused *adj*

la congestione *m* congestion *n* (sinus) p186

congestionato -a congested *adj* p186

il coniglio *m* rabbit *n*

conoscere *v* to know (someone) p20

il contachilometri *m* odometer *n* p54

i contanti *m pl* cash *n* p132

l'acconto in contanti cash advance

solo contanti cash only

il contatto d'emergenza *m*
   emergency contact n
contentissimo -a *delighted*
   adj
continuare *v* to continue **p20**
il conto *m* account *n* **p135**
il contraccettivo *m*
   contraceptive n **p191**
la contraccezione *f* birth
   control *n* **p191**
la coperta *f* blanket *n* **p47**
il coperto *m* service charge,
   cover charge *n* **p175**
coprire *v* to cover **p21**
il corno *m* horn *n*
la corona dentale *f* dental
   crown *n* **p192**
correggere *v* to correct **p20**
la corrente *f* water current *n*
correre *v* to run, to jog **p20**
corretto -a *correct adj*
il corridoio *m* hallway *n*
la corsia *f* aisle *n* (in store)
cortese *courteous adj* **p78**
 per cortesia *please*
cosa *what n* **p3**
 Cosa c'è? *What's up?*
costare *v* to cost **p20**
costoso -a *expensive adj*
il costume *m* costume *n*
 il costume da bagno *m*
   swimsuit / swim trunks n

il cotone *m* cotton *n*
la crema *f* cream *n*
 la crema solare protettiva
   sunscreen
crescere *v* to grow, to get
   larger **p32**
 Dove sei cresciuto -a?
   Where did you grow up?
il crostaceo *m* shellfish *n*
la cuccetta *f* berth *n*
la cucina *f* kitchen *n* **p73**
cucinare *v* to cook **p20**
 cotto -a *cooked adj*
 non abbastanza cotto -a
   undercooked adj
il cucinino *m* kitchenette *n*
cucire *v* to sew **p21**
il / la cugino -a *m f cousin n*
la culla *f* crib *n* **p69**
il cuoio *m* leather *n*
il cuore *m* heart *n*
il cuscino *m* pillow *n* **p47**
 il cuscino di piume *down*
   pillow

D

danneggiato -a *damaged adj*
Dannazione! *Damn! expletive*
danno *m* damage *n*
dare *v* to give **p28**
 Dia le carte anche a me.
   Deal me in.

il datore di lavoro *m*
  employer n

davanti *front adj* p42

  la porta d'ingresso *front door*

davvero *really adv*

il dazio *m duty, toll n*

  esente da tasse *duty-free*

il deambulatore *m walker n*
  (ambulatory device)

decimo -a *tenth adj* p7

del *m* / **della** *f* / **dei** *m pl* /
  delle *f pl any adj*

il delfino *m dolphin n*

la democrazia *f democracy n*

il denaro *m money n* p132

  il denaro liquido *cash*

il dente *m tooth n* p192

la dentiera *f denture n* p192

  la piastra della dentiera *denture plate*

il / la dentista *m f dentist n* p192

desiderare *v to wish* p20

derubare *v to rob* p20

  sono stato derubato. *I've been robbed.*

il / la designer *m f designer n*

la destinazione *f destination n*

la destra *f right-hand side n*

  È sulla destra. *It's on the right.*

All'angolo, girare a destra.
  Turn right at the corner.

diabetico -a *diabetic adj* p85

la diarrea *f diarrhea n* p186

Al diavolo! *Damn! expletive*

dicembre *m December n* p15

dichiarare *v to declare* **p20**

diciannove *nineteen adj* p7

diciassette *seventeen adj* p7

diciotto *eighteen adj* p7

dieci *ten adj* p7

dietro -a *behind prep adv* p5

difficile *difficult adj*

il / la dipendente *m f employee n*

dipingere *v to paint* **p20**

dire *v to say* **p32**

i diritti *m pl rights n*

  i diritti civili *civil rights*

diritto -a *straight adj adv* p5

  proprio diritto *straight ahead*

  Andare diritto. *Go straight.*
  (giving directions)

il / la disabile *m f disabled n*

il disegno *m drawing n* (art)

disponibile *available adj*

la disponibilità *f vacancy n*

distante *distant adj*

  più distante *farther*

  il / la più distante *farthest*

il distributore automatico *m vending machine n*

il **divieto** m prohibition n

Divieto di balneazione No swimming

divorziato -a divorced adj

la **doccia** f shower n p71

il **documento di riconoscimento** m identification n

dodici twelve adj p7

la **Dogana** f Customs n

il **dolce** m dessert n p98

la **lista dei dolci** dessert menu

il **dollaro** m dollar n p133

la **domanda** f question n

fare una domanda to ask a question p28

domandare v to ask (a question) p20

il **domani** m tomorrow n p4

la **domenica** f Sunday n p13

la **donna** f woman n

dopo f after, later adv, prep

dopodomani the day after tomorrow adv

doppio -a double adj

il / la **dottore -essa** m f doctor n p191

dove where adv p5

Dov'è? / Dove si trova? Where is it?

dovunque anywhere adv

la **dozzina** f dozen n

il **dramma** m drama n

due two adj p7

durare v to last p20

duro -a hard adj

il **DVD** m DVD n p157

E

è v is v. See essere **p24, 26**

ebreo -a Jewish, Hebrew adj n

Eccellente! Great! adj

l'**economia** f economy n

economico -a inexpensive adj

l'**educatore** m / l'**educatrice** f educator n p116

elaborare v to process (documents) p20

l'**elefante** m elephant n

l'**elezione** f election n p125

l'**emergenza** f emergency n

entrare v to enter p20

Proibito entrare. Do not enter.

l'**entrata** f entrance n

entusiasta enthusiastic adj

l'**erbetta** f herb n

l'**errore** m mistake n

esaurito -a sold out adj

Tutto esaurito No vacancy

l'**esca** f bait n p168

l'**esercizio fisico** m workout n

espresso -a express adj

essere *v to be* (permanent quality) **p24, 26**

C'è / Ci sono ____? *Is / Are there ____?*

l'estate *f summer n* **p15**

l'esterno *m outside n*

l'età *f age n* **p116**

l'ettaro *m hectare n* **p10**

l'etto *m a hundred grams n*

extra / in più *extra* (additional) *adj*

F

il facchino / il portiere *m porter / concierge n*

la famiglia *f family n* **p114**

fare *v to make, to do* **p28**

fare un'escursione a piedi *to hike* **p165**

Faccio un'escursione a piedi. *I'm going on a hike.*

fare il bagno *to bathe*

fare il numero *to dial* (phone) **p135**

fare il numero diretto *dial direct* **p135**

fare il surf *to surf* **p170**

fare le carte *to deal* (cards)

fare la doccia *to shower*

fare le spese *to shop*

fare l'immersione subacquea *to go scuba-diving* **p169**

fare bene *to do good*

fare male *to hurt*

Ahi! Questo fa male! *Ouch! That hurts!*

il farmaco *m medication n*

il faro della macchina *m headlight n*

fatto -a di *made of adj*

febbraio *m February n* **p14**

felice *m f happy adj* **p120**

femminile *f female adj*

la ferita *f wound, cut n* **p188**

fermare *v to stop* **p20**

Per favore si fermi. *Please stop.*

Ferma, ladro! *Stop, thief!*

la fermata *f stop n*

la fermata dell'autobus *bus stop*

la festa *f feast, holiday n*

festeggiare *v to celebrate* **p20**

il fiammifero *m match n* (stick)

il / la fidanzato -a *m f boyfriend / girlfriend n fiancé / fiancée n* **p115**

il figlio *m son n* **114**

la figlia *f daughter n* **p114**

il filo *m thread, wire n*

la finestra *f window n*

il finestrino di consegna *pickup window*

fino -a *fine adj*

il fiore *m flower n*

firmare *v to sign* **p20**

Firmi qui. *Sign here.*

il fiume *m river n*

il flauto *m flute n*

il formaggio *m cheese n*

il formato *m format n*

la formula *f formula n*

fotocopiare *v to photocopy* **p20**

fragile *fragile adj*

il / la francese *French adj n*

il francobollo *m stamp n* (postage)

il fratello *m brother n* **p115**

la freccia *f turn signal n* (car)

il freddo *m cold n* **p186**

frenare *v to brake* **p20**

il freno *m brake n* **p54**

fresco -a *fresh adj* **p109**

la fronte *f forehead n* **p118**

il frullato *m milk shake n*

la frutta *f fruit n* (collective)

i frutti di mare *m pl seafood n*

il frutto *m fruit n* **p104**

fumare *v to smoke* **p20**

la fune *f rope n*

funzionare *v to function, to work* **p20**

le funzioni religiose *f pl religious services n* **p126**

il fuoco *m fire n*

Al fuoco! *m Fire! interj*

il furgoncino *m van n* **p50**

il fusibile *m fuse n*

G

il gabbiano *m gull n*

il gabinetto *m toilet n*

la gamba *f leg n* **p190**

il gambero *m shrimp n*

il / la gatto -a *m f cat n*

il / i genitore -i *m f parent n* **p115**

gennaio *m January n* **p14**

il ghiaccio *m ice n*

la macchina per il ghiaccio *ice machine*

la giacca *f jacket n* **p152**

giallo -a *yellow adj*

il giallo *m mystery novel n*

il / la giapponese *Japanese adj n*

il / la ginecologo -a *m f gynecologist n* **p190**

giocare *v to play* (a game) **p20**

giocare a golf *to go golfing*

il giocattolo *m toy n*

il negozio di giocattoli *toy store*

il gioco *m play n*

il giornalaio *m newsstand n*

il giornale *m newspaper n*

il giorno *m day* n p161
  questi ultimi giorni *these last few days*
giovane *young adj*
il giovedì *m Thursday* n p15
girare *v to turn* **p20**
  girare a sinistra / destra *to turn left / right*
il giro turistico *m sightseeing* n
la gita *f tour* n p148
  la gita guidata *guided tour*
giù *down, downward adv*
giugno *m June* n p14
giusto -a *correct adj*
il golf *m golf* n p172
  il campo da golf *golf course*
la gomma *f rubber* n (material), *eraser, tire* n (wheel)
  Ho una gomma a terra. *I have a flat tire.*
graffiare *v to scratch* **p20**
graffiato -a *scratched adj*
il graffio *m scratch* n
il grammo *m gram* n
grande *big adj* p11
  più grande *bigger*
  il / la più grande *biggest*
il grasso *m,* **grasso -a** *fat adj n*
Grazie. *Thank you.* **interj** p1
il / la greco -a *Greek adj n*
gridare *v to shout* **p20**
grigio -a *gray adj*

la gruccia *f clothes hanger* n
il gruppo *m band* n (musical ensemble), *group* n
il guanto *m glove* n
guardare *v to look, watch* **p20**
  Guarda qui! *Look here!*
la guardia *f guard* n p37
  la guardia di sicurezza *security guard*
la guerra *f war* n p125
la guida *f guide* (publication), *tour guide* n
guidare *v to drive, guide* **p20**
il gusto *m taste* n (discernment)

H
la hostess *f flight attendant* n

I
ieri *m yesterday* n adv
  ieri l'altro *the day before yesterday*
imbarazzato -a *embarrassed adj*
l'imbarcazione *f boat* n p62
importante *important adj*
  Non è importante. *It's no big deal.*
l'imposta *f tax* n
  l'imposta sul valore aggiunto (IVA) *value-added tax (VAT)*

incassare *v to cash, to collect* (money) **p20**

incassare la vincita *to cash out (gambling)*

l'incidente *m accident n*

incinta *pregnant adj*

incominciare *v to start, to begin* **p20**

l'incrinatura *f crack n* (glass)

indicare *v to point* **p20**

l'indigestione *m indigestion n*

l'indirizzo *m address n* **p123**

Qual è l'indirizzo? *What's the address?*

indossare *v to wear* **p20**

l'indù *Hindu adj n*

l'infarto *m heart attack n*

l'infermiere -a *m f nurse n*

l'informazione *f information n*

il punto informazione *information booth*

l'ingegnere *m engineer n*

inghiottire *v to swallow* **p21**

l'inglese *m f English adj n*

l'Inghilterra *f England n*

l'ingorgo stradale *m congestion n* (traffic)

l'insalata *f salad n* **p91**

inserire la spina *v to plug in* **p21**

l'insetto *m bug, insect n*

l'insettifugo *m insect repellent n* **p187**

insultare *v to insult* **p20**

intasato -a *backed up* (toilet) *adj*

interno -a *inside adj*

l'interprete *m f interpreter n*

l'intervallo *m intermission n*

ipoudente *deaf adj* **p65**

intollerante al lattosio *lactose-intolerant adj*

l'invalidità *f disability n*

l'inverno *m winter n*

io *I pron* **p19**

l'ipovedente *m f visually-impaired n* **p65**

l'irlandese *m f Irish adj n*

l'italiano -a *m f Italian adj n*

J

il jogging *m to go running n*

K

kasher *kosher adj* **p85**

L

là *there adv* (demonstrative)

di là *over there*

la lampada *f light n* (lamp)

largo -a *wide adj*

il lato *m side n*

dall'altro lato di *across prep*

dall'altro lato della strada *across the street*

il latte *m milk n*

il latte in polvere *formula*

il lavabo *m sink n*

il lavaggio a secco *m dry cleaning n*

lavanda *f lavender n*

lavanderia a secco *f dry cleaner n* p74

lavorare *v to work* **p20**

Che lavoro fa? *What do you do for a living?*

la legge *f law n*

leggere *v to read* **p20**

leggero -a *light adj*

lei *f she, her pron* p19

lentamente *slowly adv*

le lenti a contatto *f contact lens n* p191

la lentiggine *f freckle n*

lento -a *slow adj*

le lenzuola *f sheets n* (linens)

il letto *m bed n* p67

il lettore di DVD *m DVD player n*

levare *v to take off* **p20**

la lezione *f lesson n*

libero -a *free adj*

la libreria *f bookstore n* p156

il libro *m book n* p156

il limite di velocità *m speed limit n* p56

la limousine *f limo n*

la lingua *f language n*

il liquore *m liqueur n* p89

liscio -a *smooth adj, straight adj* (drinks)

il litro *m liter n* p10

locale *local adj*

lontano -a *far adj* p5

più lontano -a *farther*

il / la più lontano -a *farthest*

loro *they, them pron* p19

la luce *f light n* p47

luglio *m July n* p15

lui *m he, him pron* p19

luminoso -a *bright adj*

luna *f moon n*

lunedì *m Monday n* p13

lungo -a *long adj* p10

più lungo -a *longer*

il / la più lungo -a *longest*

di lusso *upscale adj*

M

la macchina *f machine, car n*

la macchina cabriolet *convertible car*

la madre *f mother n* p114

maggio *m May n* (month)

la maglia *f sweater n* p152

il maiale *m pig n*

malato -a *sick adj* p48

male *badly, ache adv*

il mal d'auto *carsickness*

il mal di denti *toothache*

il mal di mare *seasickness*

il mal di testa *headache*

il / la manager *m f manager n*

la mancia *f tip n (gratuity)*

Mancia compresa. *Tip included.*

mandare *v to send* **p20**

mandare un e-mail *to send e-mail*

mandare il conto *to bill*

maneggiare *v to handle* **p20**

Maneggiare con cura. *Handle with care.*

mangiare *v to eat* **p20**

mangiare fuori *to eat out*

la mano *f sing /* **le mani** *f pl hand n* **p190**

il manuale d'istruzioni *m manual n (book)*

il marciapiede *m sidewalk n*

il marito *m husband n* **p114**

marroncino -a *tan adj*

marrone *brown adj*

il martedì *m Tuesday n* **p14**

marzo *m March n (month)*

maschile *male adj*

il maschio *m male n*

massaggiare *v to massage* **p20**

il massaggio alla schiena *m back rub n* **p179**

il mattino *m morning n*

al mattino *in the morning*

medio -a *middle, medium adj (size)* **p10**

meno *adv less* **p10**

meno *adv lesser*

il meno *n least*

il menù *m,* **la lista** *f menu n*

il menù dei bambini *children's menu*

il menù per i diabetici *diabetic menu*

il menù da asporto *takeout menu*

il mercato *m market n* **p154**

il mercatino *flea market*

il mercato all'aperto *open-air market*

il mercoledì *m Wednesday n*

il mese *m month n*

il mestiere *m occupation, trade n*

la metà *f half n*

il metro *m meter n*

il metrò *m,* **la metropolitana** *f the subway, tube n* **p63**

la linea della metropolitana *subway line*

la stazione della metropolitana *subway station*

ITALIAN—ENGLISH

Che linea di metropolitana devo prendere per____? *Which subway do I take for ____?*

mettere *v to place, to put* **p20**

mettere in borsa *to bag*

meticcio *biracial adj*

la mezzanotte *f midnight*

il mezzogiorno *m noon n*

migliore *better adj*

il migliore *best adj*

il militare *m military n*

mille *thousand adj n* **p7**

il millilitro *m milliliter n* **p10**

il millimetro *m millimeter n*

la minestra *f soup n* **p93**

il minimo *m least n*

il minuto *m minute n* **p12**

in un attimo *in a minute*

la misura *f size n* (clothing)

in modo sospetto *suspiciously adv*

la moglie *f wife n* **p115**

molti -e *many adj*

molto -a *much, very adj adv*

la moneta *f coin n* **p132**

la montagna *f mountain n*

scalare la montagna *mountain climbing*

morbido -a *soft adj*

la moschea *f mosque n* **p126**

la mostra *f exhibit n*

mostrare *v to show* **p20**

Può mostrarmelo? *Would you show it to me?*

la moto *f motorcycle n*

il motore *m engine n* **p53**

il motorino *m scooter n* **p50**

la mucca *f cow n*

la multa *f fine n* (penalty)

il museo *m museum n*

la musica *f music n* **p128**

il / la musicista *m f musician n*

il / la musulmano -a *Muslim adj n*

N

il naso *m nose n* **p118**

il nastro trasportatore *m conveyor belt n*

il naufragio *m shipwreck n*

la nausea *f nausea n* **p186**

la nave *f ship n* **p62**

la navetta *f shuttle n* (transportation)

navigare a vela *v to sail* **p20**

la nazionalità *f nationality n*

il negozio *m store, shop n*

il duty-free *duty-free shop* **p37**

il neo *m mole n* (facial feature)

il / la neonato -a *f* baby, infant *n* p116

nero -a *black adj*

nessuno -a *m f* nobody *n*

il niente *m* nothing *n*

il night *m* nightclub *n*

il / la nipote *m f* nephew, niece, grandchild *n* p115

no *no adv* p1

la nocciolina *f* peanut *n*

la noce *f* nut *n*

noi *us pron* p19

noleggiare *v* to charter (transportation), to rent p20

a noleggio *chartered, rented*

il nome *m* name *n* p114

nome e cognome *first and last name*

non-fumatori *nonsmoking adj*

la zona non-fumatori *nonsmoking area*

la carrozza non-fumatori *nonsmoking car*

la sala non-fumatori *nonsmoking room*

la nonna *f* grandmother *n*

il nonno *m* grandfather *n*, grandparent *n* p115

nono -a *ninth adj* 9

la notte *f* night *n* p13

a notte *per night*

novanta *ninety adj* p7

nove *nine adj n* p7

novembre *m* November *n* p15

nubile *f* maiden *adj n*

Ho conservato il mio nome da nubile. *I kept my maiden name.*

il numero *m* number, size *n* (shoes) p7

nuotare *v* to swim p20

la Nuova Zelanda *f* New Zealand *n*

il / la neozelandese *New Zealander adj n*

nuovo -a *new adj*

nuvoloso -a *cloudy adj* p124

O

l'oca *f* goose *n*

gli occhiali *m pl* glasses *n pl* (spectacles) p191

gli occhiali da sole *sunglasses*

l'occhio *m* eye *n* p190

occupato -a *busy adj* (phone line), taken *adj* (seat)

Questo posto è occupato? *Is this seat taken?*

offrire *v* to offer p21

l'oggi *m* today *n* p4

l'olio *m* oil *n*

l'oliva *f* olive *n*

l'ombrello *m* umbrella *n*

l'onorario *m* fee *n*

l'opera *f* opera, work *n* p148

l'operatore / l'operatrice *m f*
  phone operator *n* p135

l'optometrista *m f*
  optometrist *n*

l'ora *f* hour, time *n* p12

  A che ora? At what time?
  Che ora è? What time is it?

l'orario *m* hours (at
  museum), schedule,
  timetable *n* (train)

  È in orario? Is it on time?

ordinare *v* to order p20

ordinato -a neat *adj* (tidy)

biologico -a organic *adj*

l'organo *m* organ *n*

l'oro *m* gold *adj n*

l'orso *m* bear *n*

l'ospite *m* guest *n*

l'ostello *m* hostel *n* p66

l'otorinolaringoiatra *m*
  ear / nose / throat
  specialist *n* p188

ottanta eighty *adj* p7

l'ottavo -a eighth *adj* p9

  tre ottavi three eighths *n*

otto -a eight *adj* p7

ottobre October *n* p15

P

il pacco *m* package *n* p141

il padre *m* father *n* p115

il paesaggio *m* landscape *n*
  (painting)

pagare *v* to pay p20

  pagare il conto dell'albergo
  check out (of hotel) p78

la palestra *f* gym *n* p161

la palla *f* ball *n*

  il pallone ball, soccer (sport)

pallido -a pale *adj*

il pane *m* bread *n*

il pannolino *m* diaper *n*

  il pannolino di stoffa cloth
  diaper

il parabrezza *m* windshield *n*

la parata *f* parade *n*

parcheggiare *v* to park p20

il parcheggio *m* parking *n*

il parco *m* park *n*

il parco giochi *m* playground *n*

il parente *m* relative *n*

parlare *v* to talk, to speak
  p20

  Si parla inglese. English
  spoken here.

il parrucchiere *m* hairdresser *n*

la partenza *f* departure *n*

la partita *f* match *m* (sport)

il partito politico *m* political
  party *n*

il **passaggio pedonale** m
  walkway n
passare v to pass **p20**
il **passaporto** m passport n
il **passeggero** m passenger n
il **passeggino** m stroller n
la **passerella** f walkway,
  catwalk n

  la **passerella mobile** moving
    walkway

il **passero** m sparrow n
la **password** f password n
il **pasto** m meal n

  il **pasto per diabetici**
    diabetic meal

  il **pasto kasher** kosher meal

  il **pasto vegetariano**
    vegetarian meal

la **patente** f driver's license n
il **pavimento** m floor n
  (ground)
pazzo -a m crazy adj
il **pedaggio** m toll n
il **pediatra** m pediatrician n
pedonale pedestrian adj

  la **zona pedonale**
    pedestrian area

il **pedone** m pedestrian n
la **pelle** f skin, leather n
pentito -a sorry adj

perdere v to lose, to miss (a
  flight) **p20**

  Mi sono perso. I'm lost.

  Ho perso il passaporto.
    I lost my passport.

per favore please (polite
  entreaty) interj **p1**
il **pericolo** m danger n
la **permanente** f permanent
  (hair) n **p158**
il **permesso** m license, permit
  n **50**

  Permesso. Excuse me,
    pardon me.

permettere v to permit **p20**
il **personale** m staff n
pesare v to weigh **p20**

  Pesa ____. It weighs ____.

i **pesi** m pl weights n
il **pettine** m comb n
il **pettirosso** m robin n
il **pezzo** m / il **pezzetto** m bit
  (small amount) n
piacere v to like (take
  pleasure in) **p33**

  Mi piacerebbe ____. I
    would like ____.

il **piacere** m pleasure n

  per piacere please (polite
    entreaty)

  È un piacere. It's a pleasure.

il piano *m* floor *n*

il piano terra *ground floor*

il primo piano *first floor*

il piano tariffario *m* rate plan
n (cell phone)

il piatto *m* dish *n* p85

il piatto del giorno *special*
(featured meal)

il piccione *m* pigeon *n*

piccolo -a *little, small adj*

più piccolo -a *littler, smaller*

il più piccolo, la più piccola
littlest, smallest

il piede *m* foot (body part) *n*

pieno -a *full adj*

il pieno *full tank* (fuel)

la pila *f* battery (electric) *n*

la pillola *f* pill *n*

la pillola contro il mal di
mare *seasickness pill*

piovere *v* to rain **p20**

piovoso -a *rainy adj* p124

la piscina *f* swimming pool *n*

la pista *f* runway *n*

più tardi *later adv*

A più tardi. *See you later.*

piuttosto *rather adv*

la pizza *f* pizza *n* p92

la plastica *f* plastic *n*

lo pneumatico *m* tire *n*
(wheel) p54

poco -a *(a) little adj* (quantity)

più economico -a *cheaper*

il più economico -a
cheapest

un pó di _____
*a little bit of*_____

poco profondo -a *shallow adj*

poi *then, later adv*

la polizia *f* police *n* p192

il poliziotto *m* officer *n*

il pollame *m* poultry *n*

il pollice *m* thumb *n* p190

il pollo *m* chicken *n*

la polpetta *f* meatball *n*

le poltroncine nella sezione
orchestra *f pl* orchestra
seats *n pl*

il pomeriggio *m* afternoon *n*

nel pomeriggio / di
pomeriggio *in the afternoon*

la pompa *f* pump *n*

il ponte *m* bridge (across a
river, dental prosthesis) *n*

popolare *popular adj*

la porta *f* door *n*

la porta USB *f* USB port *n*

il portabagagli *m* trunk *n*
(car) p54

il portafogli *m* wallet *n*

portare *v* to take, to bring **p20**

portare indietro *v* to return
(something to a store) **p20**

il portiere *m goalie* (sport), *concierge n*

il porto *m port* (beverage, ship mooring) *n*

la porzione *f portion n*

la posta *f mail n* p141

la posta aerea *air mail*

la posta raccomandata *certified mail*

la posta celere *express mail*

la posta prioritaria *first class mail*

spedire per assicurata *to send by registered mail*

posta elettronica *f e-mail n*

messaggio di posta elettronica, messaggio e-mail *e-mail message*

Posso avere il suo indirizzo di posta elettronica / e-mail per favore? *May I have your e-mail address?*

il posto *m seat, place n*

il posto sul corridoio *aisle seat*

il posto in platea *orchestra seat*

Tutto a posto? *Are you okay?*

potere *v can* (be able to), *may v aux* **p30**

Posso ____? *May I ____?*

la pozza *f pool n*

il pozzo *m pit, well n*

il pranzo *m lunch, dinner, meal n*

preferire *v to prefer* **p32**

prelevare *v to withdraw* **p20**

il prelievo *m withdrawal n*

la prenotazione *m reservation n* p43

preoccupato -a *anxious adj*

preparato -a *prepared adj*

presentare *v to introduce* **p20**

Ho il piacere di presentarti ____. *I'd like to introduce you to ____.*

il preservativo *m condom n*

Hai un preservativo? *Do you have a condom?*

non senza preservativo *not without a condom*

preventivare *v to budget* **p20**

le previsioni del tempo *f weather forecast n* p125

il prezzo *m price n* p148

il prezzo del biglietto *fare / ticket price*

il prezzo d'ingresso *admission fee*

a prezzo modico *moderately priced*

ITALIAN–ENGLISH

presto -a *early, quick adj*
  È presto. *It's early.*
  Fai presto! *Be quick!*
la primavera *f spring (season) n* p15
primo -a *first adj* p9
il problema *m problem n*
il prodotto *m product n*
professionale *professional adj*
profondo -a *deep adj*
il programma *m program n*
prossimo -a *next adj*
  la prossima stazione *the next station*
il / la protestante *Protestant adj n*
provare *v to try, to try on (clothing)* p20
  provare gioia *to enjoy* p20
pulire *v to clean* p21
la pulizia *f cleanliness n*
puntare *v to bet, to put (gambling)* p20
  Punti sul rosso / nero! *Put it on red / black!*
il punteggio *m score n*
in punto *o'clock adv*
puzzare *v to smell* p20

Q

a quadretti *checked (pattern) adj*
il quadro *m painting n*

qualche *some adj* p10
qualcosa *f something n*
qualcuno -a *someone adj n*
quale *which adv* p6
qualsiasi cosa *f anything n*
quando *when adv* p4
la quantità *f amount n*
quanto *sing /* **quanti** *pl how much adj pron* p3
  Quanti? *How many?* p3
  Quanto? *How much?* p3
quaranta *forty adj* p7
quarto -a *fourth adj*
  il quarto *m quarter n*
quasi al sangue *medium rare (steak) adj*
quattordici *m fourteen adj* p7
quattro *m four adj* p7
quello -a *that adj* p19
  quelli -e *those adj*
questo -a *this adj* p19
  questi -e *these adj*
qui *here n* p5
quindici *m fifteen adj* p7
quinto -a *fifth adj* p9

R

raccomandare *v to recommend* p20
  Ti raccomando! *I beg you!*
la radio *f radio n* p51
  la radio satellitare *satellite radio*

il raffreddore *m cold n* (sickness)

il ragazzo *m boy, boyfriend n*

la ragazza *f girl, girlfriend n*

i ragazzi *m f pl kids n*

rallentare *v to slow down* **p20**

> **Rallenta!** *Slow down!*

il rame *m copper n*

la rampa *f ramp n* **p65**

il rapporto sessuale *m sexual intercourse n* **p180**

il recapito domiciliare *m home address n*

il reclamo *m complaint n*

il / la redattore -trice *m f editor n* **p121**

il regalo *m gift n*

reggere *v to hold* **p20**

restare *v to stay* **p20**

resto *m change n* (money)

la rete *f network n*

riappacificarsi *v to make up, to make peace* **p35**

riccio -a *curly adj*

il ricciolo *m curl n*

la ricerca *f search n*

la ricetta medica *f prescription n* **p188**

ricevere *v to receive* **p20**

la ricevuta *f receipt n* **p133**

richiesta *f order, request n*

riempire *v to fill* **p21**

rifiutare *v to refuse, to decline* **p20**

il rifornimento *m stock n*

rilassarsi *v to relax, to hang out* **p35**

rimediare *v to make up* **p20**

ripetere *v to repeat* **p20**

> **Può ripetere per favore?** *Would you please repeat that?*

riscuotere *v to collect* (pay) **p20**

rispondere *v to answer, to reply* **p20**

> **Per cortesia, rispondimi.** *Answer me, please.*

risposta *f answer n*

> **Ho bisogno di una risposta.** *I need an answer.*

il ristorante *m restaurant n*

il ritardo *m delay n* **p44**

> **in ritardo** *late adv*

ritirare *v to withdraw* **p20**

ritornare *v to return* (go back to) **p20**

il ritratto *m portrait n*

il ritrovo *m hangout* (hot spot) *n*

il rivelatore di metalli *m metal detector n*

la rivista *f magazine n* (periodical)

la roccia *f rock n*

  scalare le roccie *rock climbing*

romantico -a *romantic adj*

il romanzo *m novel n*

rompere *v to break* p20

rosa *pink adj*

  la rosa *f rose n*

rosso -a *red adj*

a rotelle *wheeled adj* (luggage)

la rottura *f break n*

rovesciare *v to spill* p20

rubare *v to steal, to rob* p20

  Qualcuno ha rubato il mio portafogli. *Someone stole my wallet.*

rubato -a *stolen adj* p49

il rubinetto *m faucet n*

rumoroso -a *loud, noisy adj*

la ruota di scorta *f spare tire n*

S

il sabato *m Saturday n* p14

il sacchetto *m bag n*

  il sacchetto per il mal d'aria *airsickness bag* p48

la sala d'aspetto *f lounge n*

la sala non-fumatori *f nonsmoking room n*

il saldo *m balance* (bank account) *n* p135

il sale *m salt n*

  Con poco sale *low-salt*

salire *v to climb* p21

  salire le scale *to climb the stairs*

  salire a bordo di *to board*

  Lei salirà a bordo della nave. *She will board the ship.*

salpare *v to set sail, to sail away* p20

  Quando salpiamo? *When do we sail?*

il saluto *m greeting n*

il salvagente *m life preserver n*

al sangue *rare* (meat) *adj*

sapere *v to know* (something) p20

il sapone *m soap n*

il sapore *m taste* (flavor) *n*

  il sapore di cioccolato *chocolate flavor*

il sarto *m tailor n*

il satellite *m satellite n* p51

  la radio satellitare *satellite radio*

  il tracking satellitare *satellite tracking*

sazio -a *full adj* (after a meal)

sbagliare *v to make a mistake* p20

lo sbaglio *m mistake n*

la scala *m* staircase, straight (gambling) *n*

la scala mobile *f* escalator *n*

scalare *v* to climb p20

scalare una montagna to climb a mountain

scalare le rocce rock climbing

la scalata *f* climbing *n*

l'attrezzatura per scalate climbing gear

le scale *f pl* stairs *n*

la scalinata *f* steps *n*

scaricare *v* to download, to unload p20

lo scarico *m* drain *n*

la scarpa *f* shoe *n* p151

la scheda *f* phone card *n*

la schiena *f* back *n* p190

sciolto -a loose adj

la scogliera *f* reef *n* p169

la scommessa *f* bet *n* p184

Eguaglio la tua scommessa I'll see your bet.

scommettere *v* to bet p20

scomparire *v* to disappear p21

lo sconto *m* discount *n*

lo sconto per i bambini children's discount

lo sconto per gli anziani senior discount

lo sconto per gli studenti student discount

lo scontrino *m* receipt *n*

la scottatura *f* sunburn *n*

il / la scozzese Scottish adj

lo / la scrittore -trice *m f* writer *n* p122

scrivere *v* to write p20

Come si scrive? How do you spell that?

Me lo può scrivere per favore? Would you write that down for me?

la scultura *f* sculpture *n*

la scuola *f* school *n*

la scuola media junior high / middle school

la scuola superiore high school

la facoltà di giurisprudenza law school

la facoltà di medicina medical school

la scuola elementare primary school

scusare *v* to excuse (pardon) p20

Mi scusi. Excuse me.

secco -a dry adj

secondo -a second adj p9

sedersi *v* to sit p35

la sedia a rotelle *f wheelchair n* p65

l'accesso per la sedia a rotelle *wheelchair access*

la rampa di accesso per la sedia a rotelle *wheelchair ramp*

la sedia a rotelle motorizzata *power wheelchair*

sedici *sixteen adj* p7

il sedile *m seat n*

il seggiolino *car seat, child's safety seat n*

il segnale di strada bloccata *m road-closed sign n*

il segnale di precedenza *m yield sign n*

sei *six adj* p7

self-serve *self-serve adj*

sembrare *v to look, to appear* p20

sempre *always adv*

senso unico *one way* (traffic sign) *adj* p56

il sentiero *m trail n* p165

sentire *v to hear, to feel* p21

senza *without prep*

separato -a *separated* (marital status) *adj* p116

la sera *f evening n*

di sera *at night*

la serratura *f lock n* p54

servire *v to serve* p21

servizio *m service n*

Fuori servizio *Out of service*

sessanta *sixty adj* p7

il sesso *m sex* (gender, activity) *n*

la seta *f silk n*

settanta *seventy adj* p7

sette *seven adj* p7

settembre *m September n*

la settimana *f week n* p14

questa settimana *this week*

la settimana scorsa *last week*

tra una settimana / la settimana prossima *a week from now / next week*

il settimo -a *seventh adj* p9

sgocciolare *v to drip* p20

sì *yes adv* p1

la sicurezza *f security, safety n*

il controllo di sicurezza *security checkpoint*

la guardia di sicurezza *security guard*

sicuro -a *safe* (secure) *adj*

il sigaro *m cigar n*

la sigaretta *f cigarette n*

un pacchetto di sigarette *a pack of cigarettes*

silenzioso -a *quiet adj*

simpatico -a *nice adj*

la sinfonia *f symphony n*

single *single* (unmarried) *adj*

E' single? *Are you single?*

singolo -a *single* (one) *adj*

sinistro -a *left adj* p5

a sinistra *on the left*

smarrirsi *v to get lost, to go astray* p35

smarrito -a *missing adj*

il socialismo *m socialism n*

il socio *m member n*

la soda *f seltzer n* p88

soffrire *v to suffer* p21

che soffre il mal di mare *seasick* p62

il sole *m sun n*

soleggiato -a *sunny adj* p124

solo -a *alone adj adv*

solo scotch *straight scotch*

sopra *above prep adv* p78

il sopracciglio *m eyebrow n*

la sorella *f sister n* p115

la sostituzione *f substitution n*

sottile *thin adj*

sotto *below prep, adv*

il sottotitolo *m subtitle n*

la Spagna *f Spain n*

il / la spagnolo -a *Spanish adj n*

lo spago *m twine n*

le spalle *f back n* p190

la spazzola *f brush n*

le spazzole del tergicristallo *f wiper blades n* p54

specificare *v to specify* p20

spedire *v to send, to ship* p21

le spesa *f groceries n* p100

lo spettacolo *m show n* (performance)

la spezia *f spice n*

la spia *f spy, lamp n* (dashboard) p54

la spia dell'olio *oil light*

la spiaggia *f beach n* p170

gli spiccioli *m change* (money) *n* p132

spiegare *v to explain* p20

la spina *f plug n*

spingere *v to push* p20

lo spinotto adattore *m adapter plug n* p157

sporco -a *dirty adj*

gli sport *m sports n* p163

sposarsi *v to marry* p35

sposato -a *married adj*

spuntare *v to trim* (hair) p20

lo spuntino *m snack n*

la squadra *f team n* p164

staccato -a *disconnected, detached adj*

lo stadio *m stadium n* p163

stampare *v to print* p20

stanco -a *tired, exhausted adj*

stare *v to be v* (temporary state, condition, mood) p24

stare in piedi *to stand*

lo stato *m* state *n*
  gli Stati Uniti the United States

la stazione *f* station *n*
  la stazione di polizia police station p192
  le terme spa

la stella di Natale *f* poinsettia *n*

lo / la stilista *m f* designer *n*

stitico -a constipated *adj*

lo STOP *m* STOP (traffic sign)

stordito -a dizzy *adj* p189

la storia *f* history *n*

storico -a historical *adj*

la strada *f* road, street *n*
  dall'altra parte della strada across the street
  giù per la strada down the street

stressato -a stressed *adj*

stretto -a tight, narrow *adj*

a striscie striped *adj*

su up *adv prep*

subire un furto *v* to get mugged, to get robbed p21

il succo *m* juice *n*
  il succo d'arancia orange juice
  il succo di frutta fruit juice

il sugo *m* sauce *n* p93

la suite di un albergo *f* suite *n* p67

suo -a his *adj* p19

suonare *v* to play (an instrument) p20

surriscaldare *v* to overheat p20

la sveglia *f* alarm clock *n*

la sveglia telefonica *f* wake-up call *n*

svenire *v* to faint p21

T

il tachimetro *m* speedometer *n* p54

il tacchino *m* turkey *n*

la taglia *f* size *n* (clothes)

tagliare *v* to cut p20

il taglio *m* cut *n*

il tamburo *m* drum *n*

tardi *adv* late
  Non fare tardi per favore. Please don't be late.

la targa *f* license plate *n*

la tariffa *f* fare, rate *n* (car rental, hotel) p52

la tassa *f* tax, fee *n*

il tasso d'interesse *m* interest rate *n*

la tavola di surf *f* surfboard *n*

il tavolo *m* table *n* p82

il taxi *m* taxi *n* p57
  la stazione dei taxi taxi stand

il tè *m tea n* p90
il tè con latte e zucchero *tea with milk and sugar*
il tè con il limone *tea with lemon*
la tisana *herbal tea*
il teatro *m theater n* p144
il teatro dell'opera *m opera house n* p145
il / la tedesco -a *m f German adj n*
telefonare a *v to call* (phone) p20
la telefonata *f phone call n*
una telefonata internazionale *an international phone call*
una telefonata interurbana *a long-distance phone call*
telefonico -a *telephone adj*
elenco telefonico *phone directory*
il telefonino *m cell / mobile phone n*
Vendete i telefonini con scheda prepagata? *Do you sell pre-paid phones?*
il telefono *m phone n* p135
Può darmi il suo numero di telefono? *May I have your phone number?*

la televisione *f televisi...*
la televisione via cavo ... *television*
la televisione satellitare *satellite television*
il televisore *m television set n*
il tempio *m temple n* p126
il tempo *m time n* p12
Per quanto tempo? *For how long?*
la tenda *f curtain, tent n*
tenere *v to keep, to hold* p20
tenersi per mano *to hold hands*
il tennis *m tennis n* p79
il campo da tennis *tennis court*
il tergicristallo *m windshield wiper, wiper n* p54
il terminale *m terminal n* (airport) p37
terzo -a *third adj* p9
il permesso *m permit n*
Abbiamo bisogno di un permesso? *Do we need a permit?*
il tessuto *m fabric n*
il tettuccio apribile *m sunroof n* p54
il tipo *m kind n* (type)
Che tipo è? *What kind is it?*

ITALIAN—ENGLISH

...dy out.

...gabinetto

...ere v to remove **p20**

la tonnellata f ton n

il topo m mouse n

il torto m fault n

Ho torto. I'm at fault.

Lui ha torto. He's at fault.

la tosse f cough n **p86**

tossire v to cough **p21**

il totale m total n

Qual'è il totale? What is the total?

il tovagliolo m napkin n

traboccante overflowing adj

il traffico m traffic n

Com'è il traffico? How's the traffic?

Il traffico è orribile. The traffic is terrible.

il regolamento del traffico traffic rules

la transazione f transaction n

il trasferimento m transfer n

trasferimento di valuta / di fondi money transfer

Ho bisogno di fare un trasferimento dei fondi. I need to transfer funds.

traslocare v to move **p20**

trasmettere v to transfer **p20**

tre three adj **p7**

la treccia f braid n **p158**

tredici thirteen adj **p7**

il treno m train n **p58**

trenta thirty adj **p7**

Sono le due e trenta. It's two-thirty.

il tribunale m court (legal) n

triplo -a triple adj

triste sad adj **p120**

troppo too (excessively) adv

troppo caldo -a too hot

troppo cotto -a overcooked

trovare v to find **p20**

truccarsi v to make up (apply cosmetics) **p35**

il trucco m makeup n

tu you pron (singular, informal) **p19**

tuffarsi v to dive **p35**

tuo m sing / tua f sing / tuoi m pl / **tue** f pl your adj **p19**

il tuo m / la tua f / i tuoi m pl / le tue f pl yours n **p19**

tutto- a all adj **p11**

È tutto, grazie. That's all, thank you.

tutto esaurito -a sold out adj

U

l'uccello m bird n
udire v to hear **p21**
l'ufficio postale m post office n
Dov'è l'ufficio postale?
Where is the post office?
ultimo -a last adj
umido -a humid adj **p124**
undici eleven adj **p7**
l'università f university, college n
uno -a one adj **p7**
l'uomo m man n
usare v to use **p20**
uscire v to exit **p32**
l'uscita f gate (at airport), exit n **p39**
senza sbocco not an exit
l'uscita d'emergenza emergency exit n **p41**
l'uva f grapes n

V

la vacanza f vacation, holiday n
il vagone letto m sleeping car n **p59**
la valigia f suitcase n **p49**
la valigetta f briefcase n **p49**
la vasca da bagno f bathtub n

vecchio -a old adj
vedere v to see **p22**
Posso vederlo? May I see it?
il / la vedovo -a m f widower, widow n
il vegetale m vegetable n
vegetariano -a vegetarian adj
la vela f sail n **p168**
veloce fast adj
la velocità di connessione f connection speed n **p139**
vendere v to sell **p20**
il venditore ambulante m street vendor n
venerdì m Friday n **p14**
il ventaglio m fan n
venti twenty adj **p7**
verde green adj
la verdura m vegetables n (food) **p106**
verificare v to check **p20**
la verruca f wart n
la versione f version n
il vestito m dress (garment) n
vestirsi v to dress **p35**
viaggiare v to travel, to ride **p20**
il viaggio m trip, tour n **p62**
vicino -a close, near adj, nearby adj, adv **p5**
più vicino -a closer, nearer
il / la più vicino -a closest, nearest
il / la vicino -a neighbor

il video (registratore) *m VCR,*
video n

Vietato Fumare *No Smoking*
(sign)

Vietato l'ingresso *Do not*
enter (sign)

il vigneto *m vineyard n*

il vinile *m vinyl n*

il vino *m wine n* **p88**

   il vino bianco *white wine*

   il vino rosso *red wine*

   il vino frizzante *sparkling*
   wine

   il vino dolce *sweet wine*

viola *purple adj*

la viola del pensiero *f pansy n*

il violino *m violin n*

la visione *f vision n*

visitare *v to visit* **p20**

il viso *m face n* **p118**

la vista *f view n* **p68**

   la vista sulla spiaggia *beach*
   view

   la vista sulla città *city view*

il visto *m visa n*

la vita *f life n*

il vivere *m living n*

il vocabolario *m dictionary n*

voi *you* (pl informal) *pron* **p19**

volere *v to want* **p31**

il volo *m flight n* **p38**

la volpe *f fox n*

votare *v to vote* **p20**

il voto *m grade, mark n*
(school)

W

il water *m toilet bowl n*

X

Y

Z

la zia *f aunt n* **p115**

lo zio *m uncle n* **p115**

zitto -a *quiet adj*

la zona periferica *f suburb n*

la zona non-fumatori *f non-*
smoking area n

lo zoo *m zoo n* **p130**

la zuppa *f soup n* **p93**

ITALY

MAP

NOTES (LE NOTE)

NOTES (LE NOTE)

NOTES (LE NOTE)

NOTES (LE NOTE)

NOTES (LE NOTE)

NOTES (LE NOTE)

ITALIAN—ENGLISH

NOTES (LE NOTE)

NOTES (LE NOTE)

NOTES (LE NOTE)

A Journey to the End
of the Russian Empire

Russia, 1890

ANTON CHEKHOV

A Journey to the End

of the Russian Empire

Translated by ROSAMUND BARTLETT,
ANTHONY PHILLIPS, LUBA TERPAK
and MICHAEL TERPAK

GREAT
JOURNEYS

PENGUIN BOOKS

Published by the Penguin Group
Penguin Books Ltd, 80 Strand, London WC2R ORL, England
Penguin Group (USA) Inc., 375 Hudson Street, New York, New York 10014, USA
Penguin Group (Canada), 90 Eglinton Avenue East, Suite 700, Toronto, Ontario, Canada M4P 2Y3
(a division of Pearson Penguin Canada Inc.)
Penguin Ireland, 25 St Stephen's Green, Dublin 2, Ireland (a division of Penguin Books Ltd)
Penguin Group (Australia), 250 Camberwell Road, Camberwell, Victoria 3124, Australia
(a division of Pearson Australia Group Pty Ltd)
Penguin Books India Pvt Ltd, 11 Community Centre, Panchsheel Park, New Delhi – 110 017, India
Penguin Group (NZ), 67 Apollo Drive, Rosedale, North Shore 0632, New Zealand
(a division of Pearson New Zealand Ltd)
Penguin Books (South Africa) (Pty) Ltd, 24 Sturdee Avenue, Rosebank, Johannesburg 2196, South Africa

Penguin Books Ltd, Registered Offices: 80 Strand, London WC2R ORL, England

www.penguin.com

Taken from *A Life in Letters* and *The Island: A Journey to Sakhalin*
These extracts published in Penguin Books 2007

2

A Life in Letters translation copyright © Rosamund Bartlett and Anthony Phillips, 2004
Part two reprinted with the permission of Simon and Schuster Adult Publishing Group
from *The Island: A Journey to Sakhalin* by Anton Chekhov
Copyright © 1967 by Washington Square Press
All rights reserved

The moral right of the translators has been asserted

Inside-cover maps by Jeff Edwards

Typeset by Rowland Phototypesetting Ltd, Bury St Edmunds, Suffolk
Printed in England by Clays Ltd, St Ives plc

ISBN: 978-0-141-03210-8

Contents

Anton Chekhov (1860–1904) left his comfortable and successful life in Moscow in 1890 to travel to the desolate far eastern island of Sakhalin. He described his journey to Sakhalin through the Russian Empire, across Siberia, in a series of letters from which the first part of this book is taken. The Russian government used the island of Sakhalin as a place of exile for its most dangerous prisoners. Chekhov visited the penal settlements and wrote about the island and its notorious penal colony in a powerful description of exile and isolation, sympathetically depicting humanity in the midst of the greatest physical and cultural barrenness. Chekhov's *The Island: A Journey to Sakhalin* was first published in 1893; the second part of this book is extracted from this account.

Part 1: Letters from the Eastern Empire

1. To Alexey Suvorin, 20 May 1890, Tomsk

At long last greetings from Siberian Man, dear Alexey Sergeyevich! I have been missing both you and our correspondence terribly.

I shall nevertheless start from the beginning. They told me in Tyumen that there would be no steamer to Tomsk until 18 May. I had to take horses. For the first three days every joint and tendon in my body ached, but then I got used to it and had no more pain. But as a result of the lack of sleep, the constant fussing with the baggage, the bouncing up and down and the hunger, I suffered a haemorrhage that rather spoilt my mood, which was not in any case particularly sunny. The first few days were bearable, but then a cold wind started to blow, the heavens opened and the rivers overflowed into the fields and the roads, so that I kept having to swap my vehicle for a boat. The attached pages will tell you about my battles with the floods and the mud; I didn't mention in them that the heavy boots I had bought were too tight, so that I had to plough through mud and water in felt boots which rapidly turned into jelly. The road is so bad that in the last few days I've only managed to cover forty miles or so.

[...]

I've been as hungry as a horse all the way. I filled my belly with bread in order to stop thinking of turbot, asparagus and suchlike. I even dreamt of buckwheat kasha. I dreamt of it for hours on end.

I bought some sausage for the journey in Tyumen, if you can call it a sausage! When you bit into it, the smell was just like going into a stable at the precise moment the coachmen are removing their foot bindings; when I started chewing it, my teeth felt as if they had caught hold of a dog's tail smeared with tar. Ugh! I made two attempts to eat it and then threw it away.

[...]

Oh Lord, my expenses are mounting up! Thanks to the floods I had to pay all the coachmen almost twice and sometimes three times as much as usual, for they had to work hellishly hard, it was like penal servitude. My suitcase, a nice little trunk, has proved not to be very suitable for the journey: it takes up too much room, bashes me continually in the ribs as it rattles about, and, worst of all, is threatening to fall to pieces. 'Don't take trunks on a long journey' well-meaning people told me, but I only remembered this when I had got halfway. What to do? Well, I have decided to let my trunk take up residence in Tomsk, and have bought myself some piece of shit made of leather, but which has the advantage of flopping on the floor of a tarantass[1] and adopting

---

[1] An old-fashioned four-wheeled carriage pulled by three horses, which could travel at a speed of about 8 mph.

any shape you like. It cost me 16 roubles. Anyway, to continue ... It would be sheer torture to take post-chaises all the way to the Amur. I would simply be shaken to pieces with all my belongings. I was advised to buy my own carriage. So I bought one today for 130 roubles. If I don't succeed in selling it when I get to Sretensk, where the overland part of my journey ends, I shall be left without a penny and will howl. Today I dined with Kartamyshev, the editor of *The Siberian Herald*.[2] He's a local Nozdryov,[2] a flamboyant sort of fellow ... Drank six roubles' worth.

Stop press! I have just been informed that the Assistant Chief of Police wants to see me. What can I have done?!?

False alarm. The policeman turned out to be a lover of literature and even a bit of a writer; he came to pay his respects. He's gone home to collect a play he's written; apparently he intends to entertain me with it ... He'll be back in a moment and again interrupt my writing to you ...

Write and tell me about Feodosia, about Tolstoy, the sea, the goby, the people we both know.

Greetings, Anna Ivanovna! God bless you. I often think of you.

Regards to Nastyusha and Borya. If it would give them pleasure I shall be delighted to throw myself into the jaws of a tiger and summon them to my aid, but alas! I haven't got as far as tigers yet. The only furry

---

[2] A character in Gogol's *Dead Souls*, portrayed as a drunken bully and a braggart.

animals I've seen so far in Siberia have been hundreds of hares and one mouse.

Stop press! The policeman has returned. He didn't read me his play, although he did bring it with him, but regaled me instead with a story he had written. It wasn't bad, a bit too local though. He showed me a gold ingot, and asked me if I had any vodka. I cannot recall any occasion on which a Siberian has not, on coming to see me, asked for vodka. This one told me he had got himself embroiled in a love affair with a married woman, and showed me his petition for divorce addressed to the highest authority. Thereupon he suggested a tour of Tomsk's houses of pleasure.

I've now returned from the houses of pleasure. Quite revolting. Two a.m.

What has Alexey Alexeyevich gone to Riga for? You wrote to me about this. How is his health? From now on I'll write to you punctually from every town and every station where I change horses, i.e. everywhere I have to spend the night. What a pleasure it is to have to stop for the night! Scarcely do I flop into bed but I'm asleep. Out here, when you keep going through the night without stopping, sleep becomes the most treasured prize there is; there is no greater pleasure on earth than sleep when you are tired. I now realize that in Moscow or indeed in Russia generally I have never really craved sleep. I just went to bed because it was time to do so. Not like now! Another thing I've noticed: you have no desire to drink when travelling. I haven't drunk a thing. I have smoked a lot though. I don't seem to be able to think properly; my thoughts just don't cohere. The time goes very quickly, so that you hardly notice it's moved on from ten o'clock in the morning to seven o'clock in the evening; the evening simply flows seamlessly into the morning. It's like being ill for a long time. My face is covered in fish scales because of all the wind and rain, so that when I look in the mirror I hardly recognize my former distinguished features.

I won't describe Tomsk. All Russian towns are the same. Tomsk is a dull and rather drunken sort of place; no beautiful women at all, and Asiatic lawlessness. The most notable thing about Tomsk is that governors come here to die.

I embrace you warmly. I kiss both Anna Ivanovna's hands and bow to the ground before her. It's raining. Goodbye, keep well and happy. You mustn't complain if my letters are short, slapdash or dry, because one is not always oneself while on the road and cannot write exactly as one would wish. This ink is appalling, and there always seem to be little bits of hair and other things sticking to the pen.

Your
A. Chekhov

[. . .]

2. *To Chekhov Family, 20 May 1890, Tomsk*

Dear Tungus friends! It's already Whitsuntide where you are, but here not even the willows have begun to come out and there is still snow on the banks of the Tom. I leave tomorrow for Irkutsk. I've had a rest. There was no particular point in hurrying on as the steamers across Lake Baikal don't start until 10 June, but now I'm on my way anyhow.

I am alive, in good health, and have not lost any of my money; the only problem is a slight soreness in my right eye, which is aching.

Everyone is telling me to return via America, since you apparently die of boredom sailing with the Voluntary Fleet: too much official military stuff, and you hardly ever put in to port.

Kuzovlyov, the customs officer who was exiled here from Taganrog, died two months ago in extreme poverty.

Having nothing better to do, I set down some travel impressions and sent them to *New Times*; you'll be able to read them some time after 10 June. I didn't go into anything in much detail. I was writing off the cuff – not for glory but for money, and to pay off some of the advance I received.

Tomsk is a most boring town. To judge from the drunks I have met and the supposedly intelligent people who have come to my room to pay their respects, the local inhabitants are deadly boring. At all events I find their company so disagreeable that I have given instructions that I am not receiving anyone.

I've been to the bathhouse and had some laundry done – five copecks a handkerchief! I bought some chocolate from sheer boredom.

Thanks to Ivan for the books; I can relax now. Please

send him my regards if he's not with you. I have written to Father, and would have written to Ivan except that I don't know for certain where he is living or where he has gone.

In two and a half days' time I shall be in Krasnoyarsk, and in seven and a half or eight days in Irkutsk. Irkutsk is a thousand miles from here.

I've just made some coffee and am about to drink it. It's morning, and the bells will soon start ringing for late mass.

The taiga[3] starts at Tomsk. So we'll see.

Best regards to all the Lintvaryovs and to our dear old Maryushka. I hope Mama won't worry and will pay no attention to bad dreams. Are the radishes ripe yet? There are none at all here.

Well, stay alive and healthy, and don't worry about money, there will be enough. Don't spoil the summer by skimping too much.

Your

A. Chekhov

My soul is crying out. Have mercy on me, my poor old trunk will be left behind in Tomsk and I'm buying myself a new one, soft and flat, which I can sit on, and which won't fall to bits from all the shaking around. So my poor old trunk has been condemned to end its days in exile in Siberia.

[3] Northern coniferous forest.

3. *To Chekhov Family, 25 May 1890, Mariinsk*

Spring has begun; the fields are turning green, the leaves are coming out, and there are cuckoos and even nightingales singing. It was wonderful early this morning, but at ten o'clock a cold wind started blowing, and it began to rain. It was very flat up to Tomsk, and after Tomsk it was forests, ravines and such like.

I sentenced my poor trunk to exile in Tomsk for its unwieldiness, and purchased instead (for 16 roubles!) a ridiculous object that now sprawls inelegantly on the floor of my carriage. You may now boast to everyone that we own a carriage. In Tomsk I bought a barouche with a collapsible hood etc. for 130 roubles, but needless to say it has no springs, as no one in Siberia acknowledges springs. There are no seats, but the floor is large enough and flat enough to let you stretch out full length. Travelling will be very comfortable now; I fear neither the wind nor the rain. The only thing I do fear is broken bones, because the road is truly terrible. I am endlessly getting on boats: twice this morning, and tonight we have to cover three miles by water. I am alive, and quite well.

Be well,
Your
Antoine

[...]

4. *To Alexander Chekhov, 5 June 1890, Irkutsk*

European brother!

Siberia is a big, cold country. There seems no end to the journey. There is little of novelty or interest to be seen, but I am experiencing and feeling a lot. I've battled with rivers in flood, with cold, unbelievable quagmires, hunger and lack of sleep . . . Experiences you couldn't buy in Moscow for a million roubles. You should come to Siberia! Get the courts to exile you here.

The best of the Siberian towns is Irkutsk. Tomsk is not worth a brass farthing, and none of the local districts is any better than that Krepkaya in which you were so careless as to be born. The worst of it is that in these little provincial places there is never anything to eat, and when you're on the road this becomes a matter of capital importance! You arrive in a town hungry enough to eat a mountain of food, and bang go your hopes; no sausage, no cheese, no meat, not so much as a herring, nothing but the sort of tasteless eggs and milk you find in the villages.

Generally I'm happy with my journey and glad that I came. It's hard going, but on the other hand it's a wonderful holiday. I'm enjoying my vacation.

After Irkutsk I go on to Lake Baikal, which I shall cross by steamer. Then it's 660 miles to the Amur, and from there I take a boat to the Pacific. The first thing I shall do when I get there is have a swim and eat oysters.

I arrived here yesterday and immediately made my way to the baths, then went to bed and slept. Oh, how I slept! Only now do I understand the true meaning of the word.

[. . .]

Blessings be upon you with both hands.
Your Asiatic brother A. Chekhov

5. *To Chekhov Family, 6 June 1890, Irkutsk*

Greetings, dearest Mama, Ivan, Masha and Misha and everyone. I am with you in spirit . . .

In the last long letter I wrote to you I said that the mountains round Krasnoyarsk resembled the Don ridge, but this is not really the case: looking at them from the street, I could see that they surrounded the town like high walls, and they reminded me strongly of the Caucasus. And when I left town in the early evening and crossed over the Enisei, I saw that the mountains on the far bank were really like the mountains of the Caucasus, with the same kind of smoky, dreamy quality . . . The Enisei is a wide, fast-flowing, lithe river, more beautiful than the Volga. The ferry across is wonderful, very skilfully designed in the way it goes against the current; I'll tell you more about its construction when I'm home. So the mountains and the Enisei have been the first genuinely new and original things I have encountered in Siberia. The feelings

I experienced when I saw the mountains and the Enisei paid me back a hundredfold for all the hoops I had to jump through to get here, and made me curse Levitan for being so foolish as not to come with me.

Between Krasnoyarsk and Irkutsk there is nothing but taiga. The forest is no denser than at Sokolniki, but no coachman can tell you where it ends. It seems endless; it goes on for hundreds of miles. Nobody knows who or what may be living in the taiga, but sometimes it happens in winter that people come down from the far north with their reindeer in search of bread. When you are going up a mountain and you look up and down, all you see are mountains in front of you, more mountains beyond them, and yet more mountains beyond them, and mountains on either side, all thickly covered in forest. It's actually quite frightening. That was the next new experience I had . . .

Beyond Krasnoyarsk the heat and the dust began. The heat is terrible, and I have banished my coat and hat. The dust gets into your mouth, up your nose, down your neck – ugh! To get to Irkutsk you must cross the Angara on a flat-bottomed ferry; and just then, as if on purpose, a strong wind gets up . . . I and the officers who are my travelling companions have spent the last ten days dreaming of a bath and a sleep in a proper bed, and we stand on the bank reluctantly getting used to the idea that we may have to spend the night in the village instead of in Irkutsk. The ferry simply cannot put in to shore . . . We wait an hour or two, and – oh heavens! – with a supreme effort the ferry gets to the bank and ties up alongside. Bravo, we

can have our bath, supper, and sleep. How sweet it is
to steam in the bath-house and then sleep!

Irkutsk is a splendid town, and very civilized. It has
a theatre, a museum, municipal gardens with music
playing in them, good hotels ... No ugly fences with
stupid posters and wasteland with notices saying it's
forbidden to stay there. There is an inn called the
'Taganrog'. Sugar costs 24 copecks and pine nuts are
six copecks a pound.

I was bitterly disappointed not to find a letter from
you. If you had written anything before 6 May I would
have received it in Irkutsk. I sent Suvorin a telegram
but got no reply.

Now, about sources of filthy lucre. When you need
some, write (or send a cable) to Alexander and ask him
to go to the *New Times* bookshop and collect my
royalties *for the books*. That's the first thing. The second
is, read the enclosed letter carefully and post it in
August. Keep the certificate of posting. I have written
to Alexander.

Don't forget to look out for my winning ticket.

Did I write to Misha telling him that I shall probably
come home via America? There's no need for him to
rush off to Japan.

I am alive and well, and I haven't lost any of the
money. I'm saving some of the coffee for Sakhalin. I'm
drinking excellent tea, after which I feel pleasantly
stimulated. I see a lot of Chinese. They are a good-
natured people and far from stupid. At the Bank of
Siberia I was given money straight away, received
cordially, treated to cigarettes and invited out to the

dacha. There is a wonderful patisserie, but everything is hellishly expensive. The pavements here are made of wood.

Last night the officers and I went and had a look round the town. We heard someone shouting for help about six times; it was probably somebody being strangled. We went to look, but didn't find anyone.

[...]

All my clothes are creased, dirty and torn. I look like a bandit!

[...]

6. *To Chekhov Family, 13 June 1890,* *Listvenichnaya* By the shores of Lake Baikal

I'm having the most frustrating time. On 11 June, that is the day before yesterday, we left Irkutsk in the evening, in the hope of making the Baikal steamer which was departing at four o'clock in the morning. When we reached the first station, they told us that there were no horses, so we could not continue our journey and had to spend the night there. The following morning we set out again and reached the landing stage at Lake Baikal about noon. On inquiring, we were told there would not be another steamer before Friday 15 June. So all we could do was sit on the shore until Friday, look at the water, and wait. All things eventually come to an end, and usually I don't mind waiting, except that on the 20th the steamer sails from Sretensk to go down the Amur, and if we miss that one we will have to wait until the 30th for the next steamer. God have mercy, when shall I ever get to Sakhalin?

We came to Lake Baikal by way of the banks of the Angara, which flows out of Baikal until it gets to the Enisei. Have a look at the map. The banks are very picturesque. Mountain after mountain, and all completely covered with forest. The weather has been marvellous, calm, sunny and warm; as I was travelling I felt extraordinarily well, so well in fact I find it hard to describe. It was probably due to the rest I had had in Irkutsk and to the fact that the banks of the Angara are just like Switzerland. Somehow new and original. We followed them until we came to the mouth of the river, then turned left, and there was the shore of Lake Baikal, which in Siberia they call a sea. Just like a mirror. You can't see the other shore, of course: it's more than fifty miles away. The shoreline is steep, high, rocky and tree-clad; to the right and to the left are promontories which you can see jutting out into the sea like those of Ayu-Dag or Tokhtabel near Feodosia. It's like the Crimea. The Listvenichnaya station is situated right by the water's edge and is astonishingly like Yalta; if the houses were white it would be completely like Yalta. Except that there are no buildings up on the hills: they are too sheer and it would be impossible to build on them.

We got ourselves billeted in a sort of little barn, not

unlike one of the Kraskov dachas. The Baikal starts right outside the windows, a couple of feet below the foundations. We're paying a rouble a day. Mountains, woods, the mirror-smooth Baikal – all spoilt by the knowledge that we must stay here until Friday. What are we going to do with ourselves? Also, we don't yet know what we're going to eat. The local population eats nothing but wild garlic. There's no meat or fish; despite their promises, they haven't provided any milk for us. They fleeced us of 16 copecks for a small loaf of white bread. I bought some buckwheat and a small piece of smoked ham and asked them to cook it up into a sort of mush; it tasted awful, but there was nothing we could do. One must eat. We spent all evening going round the village looking for someone to sell us a chicken, but to no avail . . . There's plenty of vodka though! Russians are such pigs. If you ask them why they don't eat meat and fish, they will tell you that there are problems with supplies and transport and so on, but you'll find as much vodka as you want even in the most remote villages. You would think it ought to be much easier to get hold of fish and meat than vodka, which costs more and is harder to transport . . . No, the point is that it is a lot more enjoyable to sit and drink vodka than make an effort to catch fish in Lake Baikal or rear cattle.

At midnight a little steamer docked; we went to have a look at it and also to find out if there might be something to eat. We were told we could have lunch the following day, but that it was night-time now, the stove in the galley was not lit, and so on. We gave

thanks for 'tomorrow' – at least there was some hope!
But alas! Just then the captain came in and announced
that the ship would be leaving at four in the morning
for Kultuk. Great! We drank a bottle of sour beer (35
copecks) in the cafeteria, which was too small even to
turn round in, and saw what looked like amber beads
sitting on a plate: this was omul caviar. We went back
to our quarters to sleep. The very idea of sleep has
become repellent to me. Each day you have to lay your
coat out on the floor wool-side up, then put your
rolled-up greatcoat and a pillow at the top, and you
sleep on these lumpy hillocks in your trousers and
waistcoat . . . Civilization, where art thou?

Today it's raining, and Baikal is shrouded in mist.
'Most diverting' as Semashko would say. It's boring. I
really should settle down to some writing, but it's hard
to work when the weather is bad. The prospect before
me is rather thankless; it would be all right if I were on
my own, but I have these officers and an army doctor
with me, and they love talking and arguing. They don't
understand much, but they talk about everything. One
of the lieutenants is also a bit of a Khlestakov show-off.
You really need to be alone when you are travelling.
It's much more interesting to sit in a coach or in your
room with your own thoughts than it is to be with
people. In addition to the army people a boy called
Innokenty Alexeyevich is travelling with us; he's a pupil
at the Irkutsk Technical School; he resembles that
Neapolitan who spoke with a lisp, but nicer and more
intelligent. We are taking him as far as Chita.

You must congratulate me: I managed to sell my

carriage in Irkutsk. I won't tell you how much I got for it, otherwise Mamasha will fall over in a dead faint and won't be able to sleep for five nights.

[. . .]

The fog has lifted. I can see clouds on the mountains. Ah, devil take it! You'd think you really were in the Caucasus . . .

Au revoir,
Your Homo Sachaliensis
A. Chekhov

7. *To Chekhov Family, 20 June 1890, Shilka, on board the steamer* Ermak

Greetings, dear household members! At last I can take off my filthy, heavy boots, my worn-out trousers, my blue shirt shiny with dust and sweat, I can wash and dress myself again like a normal human being. No more sitting in a tarantass; I am ensconced in a first-class cabin on board the Amur steamer *Ermak*. This change in my fortunes took place about ten days ago, and in the following way. I wrote to you from Listven-ichnaya to tell you about missing the Baikal steamer, which meant I would have to cross the lake not on Tuesday but on Friday, and therefore not get to the Amur steamer until 30 June. But fate can play unexpected tricks. On Thursday morning I was walking along the shore of Lake Baikal and spied smoke coming from

17

the funnel of one of the two little steamers there. Upon my asking where she was bound for, I was told she was going 'over the sea' to Klyuyevo, having been engaged by some merchant or other to take his wagon train across to the other side of the lake. Well, we also needed to go 'over the sea', and the place we needed to get to was Boyarskaya. How far was Boyarskaya from Klyuyevo? Sixteen miles, they said. I rushed off to find my travelling companions and suggested to them that we take a chance on going to Klyuyevo instead. I say 'chance', because we risked not finding any horses when we arrived at Klyuyevo, which consists of nothing but a landing-stage and a few huts, and then we would have to sit around there and miss the Friday steamer. This would have been a worse fate than the death of Igor, since that would mean having to wait until the following Tuesday. My companions nevertheless agreed to risk it, so we collected our belongings and gaily stepped on board, heading straight for the cafeteria: anything for some soup! My kingdom for a plate of soup! The cafeteria was absolutely disgusting, and built like the smallest WC you can imagine. But the cook, a former serf from Voronezh called Grigory Ivanych, proved to be a master of his craft and fed us magnificently. The weather was calm and sunny, the turquoise waters of Lake Baikal clearer than the Black Sea. People say that in the deepest places you can see down almost as far as a mile, and indeed I myself saw rocks and mountains drowning in the turquoise water that sent shivers down my spine. The trip across Baikal was wondrous, utterly unforgettable. The only bad thing was that we were in third class, the deck being fully taken up by the merchant's wagon horses, and they were stamping about during the whole voyage like raving lunatics. These horses added a certain flavour to my voyage: I felt as if I was on some sort of pirate ship. At Klyuyevo the watchman agreed to take our luggage to the station; he set off in his cart and we followed behind on foot beside the lake through the most magnificent and picturesque scenery. It's beastly of Levitan not to have come with me! The track led through the forest: woods to the right up the mountainside, woods to the left dropping down to the lake. What ravines, what crags! The colours round Baikal are warm and gentle. The weather was very warm, by the way. After a five-mile walk we came to Myskansk station, where a passing official from Kyakhta gave us excellent tea and where we managed to get horses so that we could travel on to Boyarskaya. All this meant that we left on Thursday instead of Friday, and, better still, we got away a whole twenty-four hours ahead of the mail, which usually grabs all the horses from the station. We pressed on neck and crop, nourished by the faint hope that we might reach Sretensk by the 20th. I'll wait until I see you to tell you about my journey along the banks of the Selenga and then across Trans-Baikal, except to say that the Selenga is utterly beautiful, and in Trans-Baikal I found everything I have ever wanted: the Caucasus, the Psyol valley, the area round Zvenigorod, and the Don. In the afternoon you can be galloping through the Caucasus, by nightfall you are in the Don steppe, and next morning you wake from a doze and find yourself in Poltava – and it's like that for all six hundred miles. Verkhneudinsk is a nice little town, Chita not so nice, a bit like Sumy. Needless to say, there was no time even to think of eating or sleeping; we rattled on, changing horses at stations and worrying about whether or not there would be any horses at the next one, or whether we would have to wait there for five or six hours. We covered 130 miles a day, and that's the most you can do in summer. We were completely dazed. During the day the heat was terrible too, and then it got very cold at night, so that I had to put on my leather jacket over my cotton one; one night I even had my coat on. Well, on we went and on we went, and this morning we arrived at Sretensk, exactly one hour before the steamer sailed, having tipped the coachmen at the two last stations a rouble apiece.

And so has ended my mounted journey across the wide land. It has taken me two months (I left on 21 April), and if you don't count the time spent travelling by rail and ship, the three days I was in Ekaterinburg and the week at Tomsk, a day in Krasnoyarsk and a week in Irkutsk, two days at Baikal and the days spent waiting for floods to subside so that boats could sail, you get an idea of the rapidity of my progress. I have had as good a journey as any traveller could wish for. I have not had a day's illness, and of the mass of belongings I brought with me have lost only a penknife, the strap from my trunk and a little tub of carbolic ointment. I still have all my money. Not many can travel like that for thousands of miles.

I got so used to travelling by road that now I feel

somehow ill at ease, scarcely able to believe that I am no longer in a tarantass and am not hearing the jingling of the harness bells. It is a strange feeling to be able to stretch out my legs fully when I lie down to sleep, and for my face not to be in the dust. But the strangest thing of all is that the bottle of cognac Kuvshinnikov presented me with did not get broken and has not lost a drop. I promised him that I would uncork it on the shores of the Pacific.

We are travelling down the Shilka, which joins the Argunya at Pokrovskaya station and thence flows into the Amur. It is no wider than the Psyol, perhaps not even as wide. The banks are rocky, all cliffs and forests, and full of game. We tack from side to side in order to avoid running aground or bumping the stern against the banks – the steamers and barges are always bashing alongside one another. It's very stuffy. We have just stopped at Ust-Kara, where we disembarked five or six convicts; there are mines and a hard-labour prison there.

Yesterday I was in Nerchinsk, not exactly a brilliant little place, but one could live there, I suppose.

And how are you living, ladies and gentlemen? I am completely in the dark about how things are with you. Perhaps you could club together and find a 10-copeck piece to send me a wire.

The boat is going to tie up at Gorbitsa for the night, where I shall post this letter. The nights can be misty hereabouts and it is not safe to navigate.

I am going first class, because my travelling companions are in second class and I am keen to get away from them. We were all on the road together (three in a tarantass), slept together, and have all got fed up with each other, especially I with them.

[. . .]

My handwriting is dreadfully shaky because the boat shudders all the time. It's hard to write.

A little interlude. I went down to see my lieutenants and drink tea with them. They both slept well and are in a good mood . . . One of them, Lieutenant Schmidt (not a pleasant-sounding name to my ears), from an infantry regiment, is a tall, well-fed, loud-mouthed Courlander[4] and a real Khlestakov show-off. He sings bits from all the operas but with as much ear for music as a smoked herring. He is an unhappy, rather ill-bred fellow who has squandered his entire travel allowance, knows Mickiewicz[5] by heart, is candid to a fault and can talk the hind legs off a donkey. Like Ivanenko he loves to talk about his uncles and aunts. The other officer, Meller, is a quiet, modest cartographer and a highly intelligent fellow. If it weren't for Schmidt, one could happily go a million miles with him, but when Schmidt is around shoving his oar into every conversation, I get bored with him too. Anyhow, what do you care about these lieutenants? They are not very interesting.

[4] Courland, an historic region and former duchy in Latvia, situated between the Baltic Sea and the Western Dvina River.
[5] Adam Mickiewicz (1798–1855), the Polish national poet.

Look after your health. We seem to be approaching Gorbitsa.

Hearty greetings to the Lintvaryovs. I shall write separately to Papasha. I sent a postcard to Alyosha from Irkutsk. Farewell! I wonder when this letter will get to you? Probably it will be at least forty days from now.

I embrace and bless you all. I'm missing you.

Your

A. Chekhov

[...]

8. *To Chekhov Family, 23–26 June 1890, en route from Pokrovskaya to Blagoveshchensk*

I've already written to tell you how we ran aground. At Ust-Strelka, where the Shilka flows into the Argunya (look at the map), the ship, which draws two and a half feet of water, struck a rock which holed her in a few places, and as the hold began to take in water we settled on the bottom. They set to pumping out the water and patching the holes; one of the sailors stripped naked and crawled into the hold up to his neck in water, and felt about for the holes with his heels. Each hole was then covered from the inside with heavy sail-cloth smeared with caulk, after which they placed a board over it and inserted a bracing strut on top of the board which reached up to the roof of the hold – and that was the hole repaired. They went on pumping from five o'clock in the evening until nightfall, but the water level did not go down; they had to stop work until the following morning. The next morning they discovered some more holes, so they carried on pumping and patching. The sailors pumped while we, the passengers, strolled about the decks, gossiped, ate and drank and slept; the captain and the first mate were taking their cue from the passengers and were obviously not in any hurry. To our right was the Chinese shore, to our left the village of Pokrovskaya with its Amur Cossacks. You could either be in Russia or you could cross over to China – up to you, nothing to stop you either way. During the day the heat was unbearable, and I had to put on a silk shirt.

Lunch is served at twelve noon and supper at seven o'clock in the evening.

By a piece of bad luck, the steamer coming in the opposite direction, the *Herald*, with a mass of people on board, could not get through either and both ships have ended up stuck fast. There was a military band on board the *Herald*, and the result was an excellent party; all day yesterday we had music on deck which entertained the captain and the sailors and no doubt delayed the repairs to the ship. The female passengers – particularly the college girls – were having a ball: music, officers, sailors . . . ah! Yesterday evening we went into the village, where we listened to more of the same music, which the Cossacks had paid for. Today the repairs are continuing. The captain is promising that we shall be off again after dinner, but his promises are made so languidly, his eyes wandering somewhere off to the side, that he's obviously lying. We're not in any rush. When I asked one of the passengers when we were likely to be on our way at last, he asked: 'Aren't you enjoying being here?' And he's right, of course. Why shouldn't we stay here, so long as it's not boring?

The captain, the mate and the agent are as pleasant as can be. The Chinese down in third class are good-natured, amusing people. Yesterday one of them was sitting on the deck singing something very sad in a treble voice, and while doing so his profile was more amusing than any cartoon drawing. Everyone was watching him and laughing, but he paid not the slightest attention. Then he stopped singing treble and switched to tenor – good God, what a voice! It was like a sheep bleating or a calf mooing. The Chinese remind me of gentle, tame animals. Their pigtails are long and black, like Natalia Mikhailovna's. Mention of tame animals reminds me: there is a tame fox cub living in the bathroom, which sits and watches you while you wash. If it hasn't seen anybody for a time, it starts to whimper.

We have some very strange conversations! The only topics of conversation round here are gold, the gold fields, the Voluntary Fleet, and Japan. Every peasant in Pokrovskaya, even the priest, is out prospecting for gold. So are the exiles, who can get rich here as quickly as they can get poor. There are some nouveaux riches who won't drink anything but champagne, and who will only go to the tavern if someone puts down a red carpet for them stretching right from their hut to the door of the inn.

When autumn comes, would you please send my winter coat to the *New Times* bookshop in Odessa, first asking Suvorin's permission, which you must do for form's sake. I shan't need galoshes. Also send any letters for me and a note of your address. If you should have any spare money, you could also send 100 roubles to the same address marked for transfer to me, in case I should need them. You will need to mark them to be transferred to me, otherwise I shall have to hang about getting them from the post office. If you don't have any spare, it doesn't matter. When you get to Moscow, suggest to Father that he take some potassium bromide, because he gets dizzy spells in the autumn; if this happens you must apply a leech behind his ear. Anything else? Yes, ask Ivan to buy from Ilyin (the shop in Petrovsky Lane) a map of the Trans-Baikal area printed on cloth, and send it in a printed-matter wrapper to this address: Innokenty Alexeyevich Nikitin, pupil at the Technical School. Please keep all newspapers and letters for me.

The Amur is an extraordinarily interesting and unusual region. It seethes with life in a way that you can have no conception of in Europe. It (life here, that is) reminds me of stories I've heard about life in America. The banks of the river are so wild, so unusual and so luxuriant one wants to stay here for ever. As I write these lines it is now 25 June. The steamer vibrates so much it's hard to write. We are on our way once more. I've already travelled over six hundred miles down the Amur, and have seen a million magnificent landscapes; my head is spinning with excitement and delight. I saw one cliff that would cause Kundasova to expire in ecstasy, were she to take it into her head to oxidize herself at the foot of it, and if Sofia Petrovna Kuvsh[innikova] and I were to arrange a picnic at the top of it, we could say to one another: 'You can die now, Denis, you will never write anything better.' The landscape is amazing. And it's so hot! It's warm even at night. The mornings are misty, but still warm.

I stare at the banks through binoculars and see masses of ducks, geese, divers, herons and all manner of long-billed creatures. It would be a glorious place to rent a dacha!

Yesterday we passed a little place called Reinovo, where a man in the gold business asked me to visit his sick wife. When I left his house, he pressed a wad of banknotes into my hand. I felt guilty and tried to refuse the money, handing it back and saying that I was a rich man myself; the discussion went on for some time with each of us attempting to persuade the other, but in the end I still found I had 15 roubles left in my hand. Yesterday a gold dealer with a face just like Petya Polevayev's came to lunch in my cabin; he drank champagne throughout the meal instead of water and treated us to it as well.

The villages here are like those on the Don; the buildings are a little different, but not much. The locals don't observe Lent, and they eat meat even during Passion Week; the young women smoke cigarettes and the old ones pipes – that is the custom here. It's odd to see peasant women smoking cigarettes. What liberalism! Ah, what liberalism!

The air on board gets red hot from all the talking. Out here nobody worries about saying what he thinks. There's no one to arrest you and nowhere to exile people to, so you can be as liberal as you please. The people grow ever more independent, self-sufficient and understanding. If a conflict should arise in Ust-Kara, where there are convicts working (among them many political prisoners who aren't subject to a hard-labour regime), it would spread unrest right through the whole Amur region. There's no culture of denouncing people here. A political prisoner on the run can take a steamer all the way to the ocean without fearing that the captain will turn him in. In part this can be explained by a complete indifference to what goes on in Russia. Everybody would say: 'What has that to do with me?'

I forgot to write and tell you that the coachmen in Trans-Baikal are not Russian but Buryat.[6] They are a funny lot. Their horses are viperish; they loathe being put into harness and are crazier than horses pulling fire-engines. To harness the trace horse you first have to hobble its legs; the moment the hobble is removed, the troika takes off like a bullet, enough to take your breath away. If the horse isn't hobbled it will kick over the traces and gouge chunks out of the shafts with its hooves, tear the harness to shreds and generally give an impression of a young devil caught by his horns.

We are getting near Blagoveshchensk now. Be well and happy, and don't get too used to my not being

[6] Mongolian people whose lands were located north of the Russian –Mongolian border, near Lake Baikal.

with you. But perhaps you already have? A deep bow and an affectionate kiss to you all.

Antoine

My health is excellent.

9. *To Alexey Suvorin, 27 June 1890,* *Blagoveshchensk*

Greetings, dearest friend! The Amur is a very fine river indeed; I have got from it more than I could have expected, and for some time I have been wanting to share my delight with you, but for seven days the wretched boat has been juddering so much that it has prevented me from writing. Not only that, but it is quite beyond my powers to describe the beauties of the banks of the Amur; I can but throw up my hands and confess my inadequacy. Well, how to describe them? Imagine the Suram Pass in the Caucasus moulded into the form of a river bank, and that gives you some idea of the Amur. Crags, cliffs, forests, thousands of ducks, herons and all kinds of fowl with viciously long bills, and wilderness all around. To our left the Russian shore, to our right the Chinese. If I want I can look into Russia, or into China, just as I like. China is as wild and deserted as Russia: you sometimes see villages and guard huts, but not very often. My brains have addled and turned to powder, and no wonder, Your Excellency! I've sailed more than six hundred miles down the Amur, and before that there was Baikal and Trans-Baikal ... I have truly seen such riches and experienced such rapture that death holds no more terrors for me. The people living along the Amur are most unusual, and they lead interesting lives, not at all like ours. All they talk about is gold. Gold, gold – nothing else. I feel foolish and disinclined to write, so I'm writing very briefly and like a pig; I sent you four printer's sheets today about the Enisei and the taiga, and I'll send you something later about Baikal, Trans-Baikal and the Amur. Don't throw anything away; I'll collect it all up and use it for notes to tell you in person what I seem to be unable to put on paper. I have changed ships and am now on the *Muravyov*; I'm told it is a much smoother vessel, so perhaps I shall be able to write while I'm on board.

I'm in love with the Amur and would be happy to stay here for a couple of years. It is beautiful, with vast open spaces and freedom, and it's warm. Switzerland and France have never known such freedom: the poorest exile breathes more freely on the Amur than the highest general in Russia. If you were to live here you would write a lot of splendid things that would give the public a great deal of pleasure, but I am not up to it.

Beyond Irkutsk one starts to encounter the Chinese, and by the time you get here they are more numerous than flies. They are a very good-natured people. If Nastya and Borya could get to know some Chinese, they would leave their donkeys in peace and transfer their affections to the Chinese. They are nice animals and quite tame.

The Japanese start at Blagoveshchensk, or rather

Japanese women, diminutive brunettes with big, weird hair-dos. They have beautiful figures and are, as I saw for myself, rather short in the haunch. They dress beautifully. The 'ts' sound predominates in their language. When, to satisfy your curiosity, you have intercourse with a Japanese woman, you begin to understand Skalkovsky, who is said to have had his photograph taken with a Japanese whore. The Japanese girl's room was very neat and tidy, sentimental in an Asiatic kind of way, and filled with little knick-knacks – no washbasins or objects made out of rubber or portraits of generals. There was a wide bed with a single small pillow. The pillow is for you; the Japanese girl puts a wooden support under her head in order not to spoil her coiffure. The back of her head rests on the concave part. A Japanese girl has her own concept of modesty. She keeps the light on, and if you ask her what is the Japanese word for such and such a thing she answers directly, and because she doesn't know much Russian points with her fingers or even picks it up, also she doesn't show off or affect airs and graces as Russian women do. She laughs all the time and utters a constant stream of 'ts' sounds. She has an incredible mastery of her art, so that rather than just using her body you feel as though you are taking part in an exhibition of high-level riding skill. When you climax, the Japanese girl picks a piece of cotton cloth from out of her sleeve with her teeth, catches hold of your 'old man' (remember Maria Krestovskaya?) and somewhat unexpectedly wipes you down, while the cloth tickles your tummy. And all this is done with artful coquetry, accompanied by laughing and the sing-song sound of the 'ts' . . .

When I invited a Chinaman into the cafeteria to stand him a glass of vodka, he held the glass out to me, to the barman and to the waiters before drinking it, and said 'velly nice, eat!'. That is Chinese formality. He did not drink it down in one go, as we do, but in sips, nibbling something after each sip. He then thanked me by giving me some Chinese coins. Aston-ishingly polite people! They don't spend much money on clothes, but they dress very beautifully, and they are discriminating in what they eat, which they do with a sense of ceremony.

There is no doubt that the Chinese are going to take the Amur[7] from us. Or rather, they will not take it themselves; others will take it and give it to them, the English, for example, who control China and are building strongholds everywhere. The people who live along the Amur are a very sardonic lot; they find it highly amusing that Russia is so exercised about Bul-garia, which isn't worth a brass farthing, and pays no attention whatever to the Amur. It is an improvident and foolish attitude to take. However, the politics must wait until we meet.

You sent me a telegram saying that I should make my return journey via America. I was indeed thinking of doing just that, but people are warning me against it because of the cost. There are other places besides

[7] The Amur runs along the Russo-Chinese border and its owner-ship was a constant bone of contention.

New York where you can transfer money to me; you can do so in Vladivostok, through the Bank of Siberia in Irkutsk – they welcomed me warmly when I was there. I still have some funds left, although I am spending them like water. I lost more than 160 roubles on the sale of my carriage, and my travelling companions, the lieutenants, have taken more than 100 roubles off me. But, in fact, I don't think I shall need any money transferred. If the need arises I'll let you know in good time.

I am feeling extremely well. Judge for yourself – after all, I've been living out in the open day and night for more than two months now. And all that physical exercise!

I'm rushing to get this letter finished, as the *Ermak* is due to sail back in an hour's time taking the mail with it. It will be some time in August before you receive this letter.

I kiss Anna Ivanovna's hand and pray to heaven for her good health and happiness. Has Ivan Pavlovich Kazansky been to see you, the young student with the neatly pressed trousers who makes you feel depressed?

Along the way I've done a bit of doctoring. In Rein-ovo, a little place on the Amur inhabited exclusively by gold dealers, one of them asked me to see his pregnant wife. As I was leaving his house he pressed a wad of banknotes into my hand; I felt guilty and tried to give them back, assuring him that I was a very wealthy man and didn't need the money. It ended by my giving the packet back to him, but somehow there were still 15 roubles left in my hand. Yesterday I treated a small boy and refused the six roubles his mother thrust into my hand. I'm sorry now I didn't take them.

Be well and happy. Forgive me for writing so disgracefully and with so little detail. Have you written to me in Sakhalin?

I have been swimming in the Amur. Bathing in the Amur, talking and dining with gold smugglers – is that not an interesting life?

I must run to the *Ermak*. Farewell!

Thank you for the news about my family.

Your

A. Chekhov

10. *To Chekhov Family, 29 June 1890, near Khabarovka, on board the* Muravyov

There are meteors flying all round my cabin – fireflies, just like electric sparks. Wild goats were swimming across the Amur this afternoon. The flies here are enormous. I am sharing my cabin with a Chinaman, Son-Liu-li, who chatters incessantly about how in China they cut your head off for the merest trifle. He was smoking opium yesterday, which made him rave all night and stopped me getting any sleep. On the 27th I spent some time walking round the Chinese town of Aigun. Little by little I am entering into a fantastic world. The steamer shakes so much I can hardly write. Yesterday evening I sent Papasha a congratulatory telegram. Did it arrive all right?

The Chinaman has now launched into a song inscribed on his fan. I hope you are all well.

Your

Antoine

Regards to the Lintvaryovs.

[...]

11. To Alexey Suvorin, 11 September 1890, Tatar Strait, on board the Baikal

Greetings! I'm sailing south through the Tatar Strait which separates North from South Sakhalin. I have no idea when this letter will reach you. I am in good health, although from all sides I see the green eyes of cholera staring at me, waiting to ensnare me. Cholera is everywhere – in Vladivostok, Japan, Shanghai, Chifu, Suez, even on the moon it seems – quarantine and fear are everywhere. They are expecting that cholera will strike in Sakhalin, so ships are being held in quarantine; in short, things are in a bad way. Some Europeans have died in Vladivostok, among them the wife of a general.

I stayed on North Sakhalin for exactly two months, and the local administration welcomed me there with exceptional cordiality, even though Galkin had written not a word about me. Neither Galkin, nor Baroness Muskrat nor the other geniuses I was stupid enough to turn to for help, lifted a finger to help me: I had to do everything entirely on my own account.

Kononovich, the general in charge of Sakhalin, is an intelligent and decent person. We got on well together right away, and everything turned out fine. I shall bring some papers back with me which will show you that the context in which I was working was as good as it could be. I saw *everything*, so the question now is not *what* I saw, but *how* I saw it.

I don't know exactly what will come from this, but I have achieved a good deal, enough for three dissertations. I rose every morning at five o'clock, went to bed late, and laboured all day under great pressure at the thought of how much I had still to accomplish. But now that my own experience of hard labour is over, it's hard to avoid the suspicion that in seeing all the trees I missed the wood.

By the way, I patiently carried out a census of the entire population of Sakhalin. I went to all the settlements, visited every hut and talked with everyone. I used a card system to take notes, and now have records of about ten thousand convicts and settlers. In other words, there are no convicts or settlers on Sakhalin with whom I did not meet and talk. I was especially glad to be able to make records of the children, and hope that this information will prove to be of value for the future.

I dined with Landsberg[8] and sat in the kitchen of the former Baroness Heimbruck[9]. . . . I visited all the

[8] Karl Landsberg, a guards officer exiled to Sakhalin.
[9] Baroness Olga Gembruk (Heimbruck), a convicted criminal who was exiled to Sakhalin.

celebrities. I was present at a flogging, after which I had nightmares for three or four nights about the executioner and the dreadful flogging-bench. I talked to convicts who were chained to their wheelbarrows. One day I was drinking tea in a mine, when Borodavkin, the former St Petersburg merchant who is serving a sentence here for arson, took a teaspoon out of his pocket and presented it to me. All in all it was a huge strain on my nerves and I vowed never again to come to Sakhalin.

I would like to write to you more fully, but a lady in the cabin is screaming with laughter and jabbering without ceasing, and I don't have the strength to write any more. She has been guffawing and chattering without a moment's peace since yesterday evening.

This letter will come to you via America, but I don't think I shall go that way. Everyone agrees that the route through America costs more and is more boring.

Tomorrow I shall catch a distant glimpse of the island of Matsmai, off Japan. It is now getting on for midnight, darkness is on the face of the waters and the wind is blowing. It's a mystery to me how the ship can keep going and stay on its bearings in such pitch-black conditions, not to mention in such wild and uncharted waters as the Tatar Strait.

When I remember that I am over six thousand miles away from the world I know, I feel overwhelmed with lethargy, as though it will be a hundred years before I return home.

My most profound respects and heartiest greetings

to Anna Ivanovna and all your family. May God grant you happiness and all your desires.

Your

A. Chekhov

I'm depressed.

12. To Evgenia Chekhova, 6 October 1890, South Sakhalin Island

Greetings, dear Mama! I'm writing this letter to you on what is almost the eve of my departure from here back to Russia. We wait every day for the Voluntary Fleet steamship, hoping that it will be here at the latest by 10 October. I'm sending this letter to Japan, from where it will come on to you via Shanghai, or possibly America. At present I am billeted at the Korsakovsk station, where there is no post or telegraph office, and where ships only put in once a fortnight at most. One boat did come in yesterday, bringing me a pile of letters and telegrams from the north. From them I learnt that Masha enjoyed being in the Crimea; I thought she would prefer the Caucasus. I learnt that Ivan has hopelessly failed to master the art of cooking the schoolmasterly kasha, mixing up the grains with the oats. Where is he at the moment? In Vladimir? I learnt that Mikhailo, thanks be to God, had nowhere to live all summer and so stayed at home, that you went to the Holy Mountains, and that Luka was boring and rainy. It's strange! Where you were it was rainy and cold, while from the moment I arrived in Sakhalin until today it has been warm and bright; sometimes there's a light frost in the mornings and one of the mountains has snow on the top, but the earth is still green, the leaves have not fallen and nature all around is smiling, just like May at the dacha. That's Sakhalin for you! I also found out from letters that the summer at Babkino was marvellous, that Suvorin is pleased with his house, that Nemirovich-Danchenko is not happy, that Ezhov's wife has died, poor fellow, and finally that Ivanenko and Jamais are writing to each other and that Kundasova has gone off somewhere, nobody knows where. I shall personally put Ivanenko to death, and I suppose that Kundasova is, as before, wandering the streets waving her arms about and calling everybody scum, and therefore I am not rushing to grieve for her.

At midnight yesterday I heard a ship's siren. Everyone jumped out of bed: hooray, our ship must have come in! We all got dressed, took lanterns and went down to the jetty, where indeed we saw in the distance the lights of a ship. Everyone thought it must be the *Petersburg*, the ship on which I will be sailing to Russia. I was thrilled. We climbed on board a dinghy and rowed out; we rowed, and rowed, and at last the dark bulk of the ship loomed out of the mist before us. One of us croaked out: 'Ahoy there! What ship are you?' The answer came back: *Baikal*! Oof, curses, what a disappointment! I'm homesick, and fed up with Sakhalin. After all, for three months I've seen no one besides convicts or people who have no topic of conversation other than hard labour, floggings and prisoners.

A wretched existence. I am longing to get to Japan, and then on to India.

I am very well, if you don't count a twitch in my eye which seems to be bothering me often just now, and which always seems to give me a bad headache. My eye was twitching yesterday and today, so I am writing this letter to the accompaniment of an aching head and a heaviness throughout my body. My haemorrhoids also remind me of their existence.

The Japanese Consul Kuze-San lives at Korsakovsk with his two secretaries, whom I have got to know well. They live in the European style. The local administrative establishment made an official visit today with all due pomp and circumstance, to present them with medals they had been awarded; I went along with my headache, and had to drink champagne.

While staying here in the south I went three times from the Korsakovsk station to visit Naibuchi, a place lashed by real ocean waves. Look at the map and you will find poor, benighted Naibuchi on the eastern shore of the southern island. These waves destroyed a boat with six American whalers on board; their ship was wrecked off the coast of Sakhalin and they are now living at the station, stolidly tramping the streets. They are also waiting for the *Petersburg* and will sail with me.

I sent you a letter at the beginning of September via San Francisco? Did you get it?

Greetings to Papasha, to my brothers, to Masha, to my Aunt and to Alyokha, to Maryushka, Ivanenko and all my friends. I'm not bringing any furs; there weren't any on Sakhalin. I wish you good health, and may heaven preserve you all.

Your
Anton

I'll be bringing presents for everyone. The cholera has abated in Vladivostok and Japan.

13. *To Mikhail Chekhov, 16 October 1890, Vladivostok*

Will be in Moscow on 10 December. Going via Singapore.

[. . .]

Part 2: Sakhalin Island

i. Across the Tatar Strait

Sakhalin lies in the Okhotsk Sea, protecting almost a thousand versts of eastern Siberian shoreline as well as the entrance into the mouth of the Amur from the ocean. It is long in form, running from north to south; its shape in the opinion of one author suggests a sturgeon. Its geographic location is from 45° 54' to 54° 53' latitude and from 141° 40' to 144° 53' longitude. The northern section of Sakhalin, which is crossed by a belt of permafrost, can be compared with Ryazan *guberniya*, the southern section with the Crimea. The island is 900 versts long, its widest portion measuring 125 versts and its narrowest 25 versts. It is twice as large as Greece and one and a half times the size of Denmark.

The former division of Sakhalin into northern, central and southern districts was impracticable, and it is now divided only into northern and southern. The upper third of the island precludes colonization due to its climatic and soil conditions. The central section is called Northern Sakhalin and the lower, Southern Sakhalin. There are no rigid boundaries between them. At the present time convicts inhabit the northern section along the Duyka and Tym Rivers; the Duyka falls into the Tatar Strait and the Tym into the Okhotsk

Sea; both rivers meet at their source according to the map. Convicts also live along the western bank in a small area above and below the Duyka estuary. Administratively, Northern Sakhalin is composed of two districts: Alexandrovsk and Tymovsk.

After spending the night at De Kastri, we sailed at noon on the next day, July 10, across the Tatar Strait to the mouth of the Duyka, where the Alexandrovsk command post is situated. The weather again was calm and bright, a rare phenomenon here. On the completely becalmed sea whales swam past in pairs, shooting fountains into the air. This lovely and unusual spectacle amused us the entire trip. But I must admit my spirits were depressed and the closer I got to Sakhalin the more uncomfortable I became. The officer in charge of the soldiers, learning of my mission in Sakhalin, was greatly amazed and began to argue that I had absolutely no right to visit the penal settlement and the colony since I was not a government official. Naturally I knew he was wrong. Nevertheless, I was greatly troubled by his words and feared that I would probably encounter the same point of view on Sakhalin.

When we cast anchor at nine o'clock, huge fires were burning at five different places on the Sakhalin taiga. I could not see the wharf and buildings through the darkness and the smoke drifting across the sea, and could barely distinguish dim lights at the post, two of which were red. The horrifying scene, compounded of darkness, the silhouettes of mountains, smoke, flames and fiery sparks, was fantastic. On my left monstrous fires were burning, above them the mountains, and beyond the mountains a red glow rose to the sky from remote conflagrations. It seemed that all of Sakhalin was on fire.

To the right, Cape Zhonkiyer reached out to sea, a long, heavy shoulder similar to the Crimean Ayu-Dag. A lighthouse shone brightly on the summit, while below in the water between us and the shore rose the three sharp reefs – 'The Three Brothers.' And all were covered with smoke, as in hell.

A cutter with a barge in tow approached the ship. Convicts were being brought to unload the freight. We could hear Tatar being spoken, and curses in Russian.

'Don't let them come on board,' someone shouted. 'Don't let them! At night they will steal everything on the boat.'

'Here in Alexandrovsk it is not so bad,' said the engineer, as he saw how depressed I was while gazing to shore. 'Wait until you see Dué! The cliffs are completely vertical, with dark canyons and layers of coal; fog everywhere! Sometimes we carried two to three hundred prisoners on the *Baikal* to Dué and many burst into tears when they saw the shore!'

'We are the prisoners, not the convicts,' said the captain. 'It is calm here now, but you should see it in the fall: wind, snow, storms, cold, the waves dash over the side of the ship – and that's the end of you!'

I spent the night on board. At five o'clock in the morning I was noisily awakened with, 'Hurry, hurry! The cutter is making its last trip to shore! We are leaving at once!' A moment later I was sitting in the cutter. Next to me was a young official with an angry, sleepy face. The cutter sounded its whistle and we left for the shore towing two barges full of convicts. Sleepy and exhausted by their night's labour, the prisoners were limp and sullen, completely silent. Their faces were covered with dew. I now recall several Caucasians with sharp features, wearing fur hats pulled down to their eyebrows.

'Permit me to introduce myself,' said an official. 'I am the college registrar D.'

He was my first Sakhalin acquaintance, a poet, author of a denunciatory poem entitled 'Sakhalinó,' which begins: 'Tell me, Doctor, was it not in vain . . .' Later he often visited me and accompanied me around Alexandrovsk and nearby places, relating anecdotes and endlessly reading his own compositions. During the long winter nights he writes progressive stories. On occasion he enjoys informing people that he is the college registrar and is in charge of the tenth grade. When a woman who had visited him on business called him Mr D., he was insulted and angrily screamed, 'I'm not Mr D. to you, but "your worship."' While strolling along the shore I questioned him about life on Sakhalin, about what was happening, but he only sighed ominously and said, 'You will see!'

The sun was high. Yesterday's fog and darkness, which had so terrified me, vanished in the brilliance of the early morning. The dense, clumsy Zhonkiyer with its lighthouse, 'The Three Brothers' and the high, craggy shores which were visible for tens of versts on both sides, the transparent mist on the mountains and the smoke from the fires did not present such a horrifying scene in the bright sunlight.

There is no harbour here, and the coast is dangerous. This fact was impressively demonstrated by the presence of the Swedish ship *Atlas*, which was wrecked shortly before my arrival and now lay broken on the shore. Boats usually anchor a verst from shore and rarely any nearer. There is a pier, but it is only usable by cutters and barges. It is a large pier, several sazhens long, and T-shaped. Thick log piles had been securely driven into the sea bottom, in the form of squares, which were filled with stone. The top was covered with planking, and there were freight-car rails running the length of the pier. A charming building, the pier office, sits on the wide end of the T; here also stands a tall black mast. The construction is solid, but not permanent. I was told that during a heavy storm the waves sometimes reach the windows of the building and the spray even reaches the yardarm of the mast; the entire pier trembles.

Along the shore near the pier some 50 convicts were wandering, obviously idle; some were in overalls, others in jackets or grey cloth coats. When I approached, they all removed their caps. It is possible that no writer has ever previously received such an honour. Somebody's horse was standing on shore harnessed to a springless carriage. The convicts loaded my luggage in the carriage; a black-bearded man in a coat with his shirt tail hanging got up on the box. We took off.

'Where do you wish to go, your worship?' he asked, turning around and removing his cap.

I asked him if it would be possible to rent lodgings here, even if it was only one room.

'Certainly, your worship, rooms can be rented.'

For the two versts from the pier to the Alexandrovsk Post I travelled along an excellent highway. In comparison to the Siberian roads this is a clean, smooth road with gutters and street lights; it is absolutely luxurious. Adjacent to it runs a railway. However, the scenery along the way is depressing in its barrenness. Along the tops of the mountains and hills encircling the Alexandrovsk valley, through which the River Duyka flows, charred stumps and trunks of larch trees, dried out by fire and wind, project like porcupine quills, while in the valley below there are hillocks covered with sorrel – the remains of swamps which until recently were impassable. The fresh slashes in the earth made by the gutters reveal the complete barrenness of the swampy scorched earth with its half-*vershok* layer of poor soil. There are no spruce trees, no oaks, no maples – only larches, gaunt, pitiful, fretted in precise shapes, and they do not beautify the forests and parklands as they do in Russia, but serve only to emphasize the poor marshy soil and the severe climate.

The Alexandrovsk Post, or Alexandrovsk for short, is a small, pretty Siberian-type town with 3,000 inhabitants. It does not contain even one stone building. Everything is built of wood, chiefly of larch – the church, the houses and the sidewalks. Here is located the residence of the island's commandant, the centre of Sakhalin civilization. The prison is situated near the main street. Its exterior is quite similar to an army barracks, and as a result Alexandrovsk is completely free of the dismal prison atmosphere which I had expected.

The driver took me to the Alexandrovsk residential district in the suburbs, to the home of one of the peasant exiles. Here I was shown my lodgings. There was a small yard, paved Siberian fashion with timbers and surrounded with awnings. The house contained five spacious, clean rooms and a kitchen, but not a stick of furniture. The landlady, a young peasant woman, brought out a table, and a chair came about five minutes later. 'With firewood the price is 25 roubles; without firewood, 15,' she said.

About an hour later, she brought a samovar and said with a sigh:

'So you have come to visit this godforsaken hole!'

She had come as a little child with her mother, following her father, a convict who has not yet served out his sentence. Now she is married to one of the exiled peasants, a gloomy old man whom I glimpsed crossing the yard. He had some sort of sickness and spent his time lying under the awning and groaning.

'At home in Tambovsk *guberniya* they are probably reaping,' she said. 'Here there is nothing to look at.'

And truly there is nothing interesting to look at. Through the window you could see rows of cabbage plants, and some ugly ditches nearby, and beyond these a gaunt larch tree withering away.

Groaning and holding his side, the landlord entered and began complaining of crop failure, the cold climate, the poor soil. He had completed his prison term and exile, and now owned two houses, some horses and a cow. He employed many workmen and did nothing himself. He had married a young woman and, most important, he had long since been granted permission to return to the mainland – but still he complained.

[. . .]

The prisoners and the exiles, with some exceptions, walk the streets freely, without chains and without guards; you meet them in groups and singly every step of the way. They are everywhere, in the streets and in the houses. They serve as drivers, watchmen, chefs, cooks and nursemaids. I was not accustomed to seeing so many convicts, and at first their proximity was disturbing and perplexing. You walk past a construction site and you see convicts with axes, saws and hammers. 'Well,' you think, 'they are going to haul me off and murder me!' Or else you are visiting an acquaintance and, not finding him at home, you sit down to write a note, while his convict servant stands waiting behind you, holding the knife with which he has been peeling potatoes in the kitchen. Or it may happen that at about four o'clock in the morning you will wake up and hear a rustling sound, and you look and see a convict approaching the bed on tiptoe, scarcely breathing.

'What's the matter? What do you want?'

'To clean your shoes, your worship.'

Soon I became accustomed to this. Everyone becomes accustomed to it, even women and children. The local ladies think nothing of permitting their

49

children to go out and play in the care of nursemaids sentenced to exile for life.

One correspondent writes that at first he was terrified of every bush, and groped for the revolver under his coat at every encounter with a prisoner on the roads and pathways. Later he calmed down, having come to the conclusion that 'the prisoners are generally nothing more than a herd of sheep, cowardly, lazy, half-starved and servile.' To believe that Russian prisoners do not murder and rob a passerby merely out of cowardice and laziness, one must be either a very poor judge of men or not know them at all.

[. . .]

The days were beautiful with a bright sky and clear air, reminiscent of fall in Russia. The evenings were magnificent. I remember the glowing western sky, the dark-blue sea and a completely white moon rising over the mountains. On such evenings I enjoyed driving along the valley between the post and the village of Novo-Mikhaylovka; the road is smooth, straight; alongside is a railway and a telegraph line. The further we drove from Alexandrovsk, the more the valley narrowed, the shadows deepened; there were giant burdocks in tropical luxuriance; dark mountains rose on all sides. In the distance we could see the flames from coke fires, and there were more flames from a forest fire. The moon rose. Suddenly a fantastic scene. Coming toward us along the railway was a convict, riding in a small cart, dressed in white and leaning on a pole. He stopped abruptly.

'Isn't it time to turn back?' asked my convict driver.

Then he turned the horses, and glancing up at the mountains and the fires, he said:

'It is lonesome here, your worship. It is much better at home in Russia.'

ii. The Prison Settlements of Northern Sakhalin

The second district of Northern Sakhalin is located on the other side of a ridge of the mountain range and is called Tymovsk, because its settlements lie along the Tym River, which falls into the Okhotsk Sea. As you drive from Alexandrovsk to Novo-Mikhaylovka, the mountain ridge rises before you and blocks out the horizon, and what you see from there is called the Pilinga. From the top of the Pilinga a magnificent panorama opens out with the Duyka valley and the sea on one side, and on the other a vast plain which is watered by the Tym and its tributaries for more than 200 versts. This plain is far more interesting than Alexandrovsk. The water, the many kinds of timber forests, the grasses which grow higher than a man, the fabulous abundance of fish and coal deposits suggest the possibility of a satisfying and pleasant life for a million people. That is the way it should be, but the frozen currents of the Okhotsk Sea and the ice floes floating on the eastern shore even in June attest with incontrovertible clarity to the fact that when nature created Sakhalin man and his welfare was the last thing in her mind. If it were not for the mountains, the plain would be a tundra, colder and bleaker than around Viakhty.

The first person to visit the Tym River and describe it was Lieutenant Boshnyak. In 1852 he was sent here by Nevelskoy to verify information obtained from Gilyaks about coal deposits and to cross the island all the way to the shore of the Okhotsk Sea, where there was said to be a beautiful harbour. He was given a dog team, hardtack for thirty-five days, tea and sugar, a small hand compass and a cross. With these came Nevelskoy's parting words of encouragement: 'As long as you have hardtack to quieten your hunger and a mug of water to drink, then with God's help you will find it possible to do your job.'

Having made his way down the Tym to the eastern shore and back, he somehow reached the western shore, completely worn out and famished, and with abscesses on his legs. The starving dogs refused to go any farther. He spent Easter day huddled in the corner of a Gilyak yurt, utterly exhausted. His hardtack was gone, he could not communicate with the Gilyaks, his legs were giving him agonies of pain. What was most interesting about Boshnyak's explorations was, quite obviously, the explorer himself, his youth – he was only twenty-one years old – and his supremely heroic devotion to his task. At the time the Tym was covered with deep snow, for it was March . . .

[. . .]

In 1881 the zoologist Polyakov carried out some serious and extensive explorations of the Tym from a scientific and practical point of view. He left Alexandrovsk on July 24, driving oxen, and crossed the Pilinga with the greatest difficulty. There were only footpaths, and these were climbed by convicts carrying provisions on their backs from the Alexandrovsk district to the Tymovsk.

The elevation of the ridge is 2,000 feet. On a Tym tributary, the Admvo, close to the Pilinga, stood the Vedernikovsky way station, of which only one position has survived, the office of the station guard.

The Tym tributaries are fast flowing, tortuous and full of rapids. It is impossible to use boats. Therefore it was necessary for Polyakov to go by oxen to the Tym River. From Derbinskoye he and his companion used a boat throughout the whole length of the river.

It is tiresome to read his account of this journey because of the exactitude with which he recorded all the rapids and sandbanks. In the 272 versts from Derbinskoye to the sea he was forced to overcome 110 obstacles: 11 rapids, 89 sandbanks and 10 places where the water was dammed by drifting trees and bushes. This means that on the average of every two miles the river is either shallow or choked up. Near Derbinskoye it is 20–25 sazhens wide: the wider the river, the shallower. The frequent bends and turns, the rapid flow and the shallows offer no hope that it will ever be navigable in the real sense of the word. In Polyakov's opinion it would probably be used only for floating rafts. Only the last 70 to 100 versts from the mouth of the river, where it is least favourable for colonization, are deeper and straighter. Here the flow is slower, and there are no rapids or sandbanks. A steam cutter or even a shallow-draft tugboat could use this part of the river.

When the rich fisheries in the neighbourhood fall into the hands of capitalists, serious attempts will probably be made to clear and deepen the waterway. Perhaps a railroad will be built along the river to its mouth, and there is no doubt that the river will repay all these expenditures with interest. But this is far in the future. Under existing conditions, when we consider only the immediate future, the riches of the Tym are almost an illusion. It offers disappointingly little to the penal colony. The Tymovsk settler lives under the same starvation conditions as the Alexandrovsk settler.

According to Polyakov, the Tym River valley is dotted with lakes, bogs, ravines and pits. It has no straight and level expanses overgrown with nutritious fodder grasses, it has no fertile meadows watered by spring floods, and only rarely are sedge-covered meadows found – these are islands overgrown with coarse grass. A thick coniferous forest covers the slopes of the hill. On these slopes we find birches, willows, elms, aspens and entire stands of poplars. The poplars are extremely tall. They are undermined at the banks and fall into the water, where they look like bushes and beaver dams. The bushes here are the bird cherry, the osier, the sweetbrier, the hawthorn . . . Swarms of mosquitoes are everywhere. There was frost on the morning of August 1.

The closer you get to the sea, the sparser the vegetation. Slowly the poplar vanishes, the willow tree becomes a bush; the general scene is dominated by the sandy or turfy shore with whortleberries, cloudberries and moss. Gradually the river widens to 75–100 sazhens; now the tundra has taken over, the coastline consists of lowlands and marshes . . . A freezing wind blows in from the ocean.

The Tym falls into Nyisky Bay, or the Tro, a small watery wasteland which is the doorway to the Okhotsk Sea, or, which is the same thing, into the Pacific Ocean. The first night Polyakov spent on the shores of the bay was bright and chilly, and a small twin-tailed comet glistened in the sky. Polyakov does not describe the thoughts which crowded in upon him as he enjoyed the sight of the comet and listened to the sounds of the night. Sleep overtook him. On the next day fate rewarded him with an unexpected spectacle. At the mouth of the bay stood a dark ship with some white strakes; the rigging and deckhouse were beautiful; a tied live eagle sat on the prow.

The shore of the bay made a dismal impression on Polyakov. He calls it a typically characteristic example of a polar landscape. The vegetation is meagre and malformed. The bay is separated from the sea by a long, narrow sandy tongue of land created by dunes, and beyond this slip of land the morose, angry sea has spread itself boundlessly for thousands of versts. When a little boy has been reading Mayne Reid and his blanket falls off during the night, he starts shivering, and it is then that he dreams of such a sea. It is a nightmare! The surface is leaden, over it there hangs a monotonous grey sky, and the savage waves batter the wild treeless shore. The waves roar, and once in a great while the black shape of a whale or a seal flashes through them.

Today there is no need to cross the Pilinga by climbing over steep hills and through gulleys in order to reach the Tymovsk district. I have already stated that people nowadays travel from Alexandrovsk to the Tymovsk district through the Arkovo valley and change horses at the Arkovo way station. The roads here are excellent and the horses can travel swiftly.

The first settlement of the Tymovsk region lies sixteen miles past the Arkovo way station bearing the Oriental fairy-tale name of Upper Armudan. It was founded in 1884 and consists of two parts which have spread along the slopes of the mountain near the Armudan River, a tributary of the Tym. It has 178 inhabitants: 123 male and 55 female. There are 78 homesteads with 28 co-owners. Settler Vasilyev even has two co-owners. In comparison with Alexandrovsk, the majority of the Tymovsk settlements, as the reader will see, have many co-owners or half-owners, few women and very few legally married families. In Upper Armudan, of 48 families, only 9 are legal. There are only three free women who followed their husbands, and it is the same in Krasny Yar or Butakovo, which are no more than a year old. This insufficiency of women and families in the Tymovsk settlements is often astounding, and does not conform with the average number of women and families on Sakhalin. It cannot be explained by any local or economic conditions, but by the fact that newly arrived prison parties are sorted out in Alexandrovsk, and the local authorities, according to the proverb that 'your own shirt is nearest to your body,' retain the majority of the women in their own district and 'keep the best for themselves; the worst they send to us,' as a Tymovsk official told me.

The huts in Upper Armudan are either thatched or

covered with tree bark; some windows have no panes or are completely boarded up. The poverty is terrible. Twenty of the men do not live at home. They have gone elsewhere to earn a livelihood. Only 60 desyatins of land have been cultivated for all 75 homesteads and 28 co-owners; 183 poods of grain have been sown, which is less than 2 poods per household. It is beyond my understanding how grain can be grown here, however much is sown. The settlement is high above sea level and is not protected from northern winds; the snow melts two weeks later than in the neighbouring settlement of Malo-Tymovo. In order to fish, they travel 20 to 25 versts to the Tym River in the summer. They hunt fur animals more for sport than for gain, and so little accrues to the economy of the settlement that it is scarcely worth talking about.

I found the householders and the members of their households at home; none of them were occupied even though it was not a holiday, and it seemed that during the warm August weather all of them, from the youngest to the oldest, could have found work either in the field or on the Tym, where the periodic fish were running. The householders and their cohabitants were obviously bored and eager to sit down and discuss anything at all. They laughed from boredom and sometimes cried. They are failures, and most of them are neurasthenics and whiners, 'alienated persons.' Forced idleness has slowly become a habit and they spend their time waiting for good sea weather, become fatigued, have no desire to sleep, do nothing, and are probably no longer capable of doing anything except shuffling cards. It is not strange that card-playing flourishes in Upper Armudan and the local players are famous all over Sakhalin. Because of lack of money they play for small stakes, but make up for this by playing continually, as in the play *Thirty Years, or the Life of a Card Player*. I had a conversation with one of the most impassioned and indefatigable card-players, a settler called Sizov:

'Your worship, why don't they send us to the mainland?' he asked.

'Why do you want to go there?' I asked jokingly. 'You'll have no one to play cards with.'

'That's where the real games are.'

'Do you play faro?' I asked, and held my tongue.

'That's right, your worship, I play faro.'

Later, upon leaving Upper Armudan, I asked my convict driver:

'Do they play for winnings?'

'Naturally, for winnings.'

'But what do they lose?'

'What do you mean? Why, they lose their government rations, their smoked fish! They lose their food and clothing and sit about in hunger and cold.'

'And what do they eat?'

'Why, sir, when they win, they eat; when they lose, they go to sleep hungry.'

Along the lower reaches of the same tributary there is a smaller settlement, Lower Armudan. I arrived late at night and slept in a garret in the jail because the jailer did not permit me to stay in a room. 'It's impossible to sleep here, your worship; the bugs and cockroaches win all the time!' he said helplessly, spreading his hands wide. 'Please go up to the tower.' I climbed to the tower on a ladder, which was soaked and slippery from the rain. When I descended to get some tobacco I saw the 'winning creatures,' and such things are perhaps only possible on Sakhalin. It seemed as though the walls and ceiling were covered with black crêpe, which stirred as if blown by a wind. From the rapid and disorderly movements of portions of the crêpe you could guess the composition of this boiling, seething mass. You could hear rustling and a loud whispering, as if the insects were hurrying off somewhere and carrying on a conversation.

There are 101 settlers in Lower Armudan: 76 male and 25 female. There are 47 homesteaders with 23 co-owners. Four families are married; 15 live as cohabitants. There are only two free women. There are no inhabitants between 15 and 20 years of age. The people live in dire poverty. Only six of the houses are covered with planking; the rest are covered with tree bark and, as in Upper Armudan, some have no windowpanes or are boarded up. My records include not a single labourer. Obviously the householders do nothing. In order to find work, 21 of them have left. Since 1884, when the settlement was founded, only 37 desyatins of arable land have been cleared – i.e., one-half desyatin per homestead. One hundred and eighty-three poods of winter grain and summer corn have been sown. The settlement in no way resembles an agricultural village. The local inhabitants are a disorganized rabble of Russians, Poles, Finns and Georgians, starving and ragged, who came together not of their own volition but by chance, after a shipwreck.

The next settlement along the route lies on the Tym. Founded in 1880, it was named Derbinskoye in honour of the jailer Derbin, who was murdered for his cruelty. He was still young, but a brutish, stern and implacable fellow. The people who knew him recall that he always walked around the prison and on the streets with a stick which he used for beating people. He was murdered in the bakery. He defended himself and fell into the fermenting bread batter, bloodying the dough. His death was greeted with great rejoicing by the convicts, who donated a purse of 60 roubles to the murderer.

There is nothing else amusing in Derbinskoye. It lies on a flat and narrow piece of land, once covered with a thick birch and ash forest. Below, there is a wide stretch of marshland, seemingly unfit for settlement, once thickly covered with fir and deciduous trees. They had scarcely finished cutting down the forest and clearing stumps in order to build the huts, the jail and the government storehouse, and draining the area, when they were forced to battle with a disaster which none of the colonizers had foreseen. During the spring, the high water of the Amga stream flooded the entire settlement. They had to dig another bed and rechannel it. Now Derbinskoye has an area of more than a square verst and resembles a real Russian village.

You enter by a splendid wooden bridge; the stream babbles, the banks are green with willows, the streets are wide, the huts have plank roofs and gardens. There are new prison buildings, all kinds of storehouses and warehouses, and the house of the prison warden stands in the middle of the settlement, reminding you not so much of a prison as of a manorial estate. The warden is continually going from warehouse to warehouse, and he clanks his keys exactly like a landlord in the good old days who guards his stores day and night. His wife sits near the house in the front garden, majestic as a queen, and she sees that order is kept. Right in front of her house, in an open hothouse, she can see her fully ripened watermelons. The convict gardener Karatayev tends them with indulgence and with a slavish diligence. She can see the convicts fishing in the river, bringing back healthy, choice salmon called *serebryanka* [silver fish], which are then cured and given to the officials; they are not given to the convicts. Near the garden play little girls dressed like angels. A convict dressmaker, convicted for arson, sews their clothes. There is a feeling of quiet contentment and ease. These people walk softly like cats, and they also express themselves softly, in diminutives: little fish, little cured fish, little prison rations . . .

There are 739 inhabitants in Derbinskoye, 442 male and 297 female. Altogether, including the prison population, there is a total of about 1,000. There are 250 householders and 58 co-owners. In its outward aspects as well as in the age groups of the inhabitants and, generally, in all the statistics concerning the place, it is one of the few settlements on Sakhalin which can seriously be called a settlement and not a haphazard rabble of people. It has 121 legitimate families. Twelve of them are free, and among the legally married, free women predominate. There are 103 free women. Children comprise one-third of the population.

However, in attempting to understand the economic status of the Derbinskoye inhabitants, you have to confront the various chance circumstances, which play their major and minor roles as they do in other Sakhalin settlements. Here natural law and economic laws appear to take second place, ceding their priority to such accidental variables as the greater or lesser number of unemployables, the number of sick people, the number of robbers, the number of former citizens forced to become farmers, the number of old people, their proximity to the prison, the personality of the warden, etc., etc., and all of these conditions can change every five years or even less than five years. Those who completed their sentences prior to 1881 were the first to settle here, carrying on their backs the bitter past of the settlement, and they suffered, and gradually took over the better land and homesteads. Those who arrived from Russia with money and families are able to live well. The 220 desyatins of land and the yearly production of 3,000 poods of fish, as shown in the records, obviously pertain to the economic position of these homesteaders. The remainder of the inhabitants, more than one-half of the population of Derbinskoye, are starving, in rags, and give the impression of being useless and superfluous; they are hardly alive, and they prevent others from living. In our own Russian villages even fires produce no such sharp distinctions.

It was raining, cold and muddy when I arrived in Derbinskoye and visited the huts. Because of his own

Anton Chekhov

small quarters, the warden gave me lodging in a new, recently completed warehouse, which was stored with Viennese furniture. They gave me a bed and a table, and put a latch on the door so that I could lock myself in from inside.

All evening to two o'clock in the morning I read or copied data from the list of homesteads and the alphabetical list of the inhabitants. The rain fell continually, rattling on the roof, and once in a while a belated prisoner or soldier passed by, slopping through the mud. It was quiet in the warehouse and in my soul, but I had scarcely put out the candle and gone to bed when I heard a rustling, whispering, knocking, splashing sound, and deep sighs. Raindrops fell from the ceiling on to the latticework of the Viennese chairs and made a hollow, ringing sound, and after each such sound someone whispered in despair: 'Oh, my God, my God!' Next to the warehouse was the prison. Were the convicts coming at me through an underground passage? But then there came a gust of wind, the rain rattled even more strongly, somewhere a tree rustled – and again, a deep, despairing sigh: 'Oh, my God, my God!'

In the morning I went out on the steps. The sky was grey and overcast, the rain continued to fall, and it was muddy. The warden walked hurriedly from door to door with his keys.

'I'll give you such a ticket you'll be scratching yourself for a week,' he shouted. 'I'll show you what kind of ticket you'll get!'

These words were intended for a group of twenty

prisoners who, from the few phrases I overheard, were pleading to be sent to the hospital. They were ragged, soaked by the rain, covered with mud and shivering. They wanted to demonstrate in mime exactly what ailed them, but on their pinched, frozen faces it somehow came out false and crooked, although they were probably not lying at all. 'Oh, my God, my God!' someone sighed, and my nightmare seemed to be continuing. The word 'pariah' comes to mind, meaning that a person can fall no lower. During my entire sojourn on Sakhalin only in the settlers' barracks near the mine and here, in Derbinskoye, on that rainy, muddy morning, did I live through moments when I felt that I saw before me the extreme limits of man's degradation, lower than which he cannot go.

In Derbinskoye there is a convict, a former baroness, whom the local women call 'the working baroness.' She lives a simple, labourer's life, and they say she is content with her circumstances. One former Moscow merchant who once had a shop on Tverskaya-Yamskaya told me with a sigh, 'The racing season is on in Moscow,' and then, turning to the settlers, he began to explain what kind of races they were and how many people go on Sundays to the racecourse along Tverskaya-Yamskaya. 'Believe me, your worship,' he said, his excitement mounting as he discussed the racecourse, 'I would give everything, my whole life, if I could see not Russia, not Moscow, but the Tverskaya!'

In Derbinskoye there live two people called Emelyan Samokhvalov, who are related to one another, and I remember that in the yard of one of them I saw a rooster tied up by its legs. The people of Derbinskoye are amused by the fact that these two Emelyan Samokhvalovs were by a strange and very complex combination of events brought together from the opposite ends of Russia to Derbinskoye, bearing the same name and being related to one another.

On August 27, General Kononovich arrived in Derbinskoye with the commandant of the Tymovsk district, A. M. Butakov, and another young official. All three were intelligent and interesting people. The four of us went on a small trip. From beginning to end we were beset with so much discomfort that it turned out to be not a trip at all; it was a parody of an expedition.

First of all, it was pouring. It was muddy and slippery; everything you touched was soaking wet. Water leaked through our collars after running down our necks; our boots were cold and wet. To smoke a cigarette was a complicated, difficult affair which was accomplished only when we all helped one another. Near Derbinskoye we got into a rowboat and went down the Tym. On the way we stopped to inspect the fisheries, a water mill and ploughland belonging to the prison. I will describe the fishing elsewhere; we all agreed the water mill was wonderful; and the fields were nothing special, being interesting only because they were so small; a serious homesteader would regard them as child's play.

The river was swift, and the four rowers and the steersman worked in unison. Because of the speed and frequent bends in the river, the scenery changed every minute. We were floating along a mountain taiga river, but all of its wild charms, the green banks, the steep hills and the lone motionless figures of the fishermen, I would have enthusiastically exchanged for a warm room and dry shoes, especially since the landscape was monotonous, not novel to me, and, furthermore, it was covered with grey, rainy mist. A. M. Butakov sat on the bow with a rifle and shot at wild ducks which were startled at our approach.

Northeast from Derbinskoye along the Tym there are only two settlements to date: Voskresenskoye and Uskovo. To settle the Tym up to its mouth would require at least thirty such settlements with ten versts between each of them. The administration plans to set up one or two every year, connecting them with a road which will eventually span the distance between Derbinskoye and Nyisky Bay. The road will bring life and stand guard over a whole series of settlements. As we came close to Voskresenskoye, a guard stood at attention, obviously expecting us. A. M. Butakov shouted to him that on returning from Uskovo we would spend the night there and that he should prepare more straw.

A little while later, the air was strongly permeated with the stench of rotting fish. We were approaching the Gilyak village of Usk-vo, the former name of the present Uskovo. We were met on shore by Gilyaks, their wives, children and bobtailed dogs, but our coming was not regarded with the same amazement as the coming of the late Polyakov. Even the children and the dogs looked at us calmly.

Anton Chekhov

The Russian colony is two versts from the riverbank. In Uskovo the same conditions exist as in Krasny Yar. The street is wide with many tree trunks still to be uprooted, full of hillocks, covered with forest grass, and on each side stand unfinished huts, felled trees and piles of rubble. All new construction on Sakhalin gives the impression of having been destroyed by an enemy or else of being long since abandoned. Only the fresh, bright colours of the hut frames and the shavings give evidence that something quite opposite to destruction is taking place.

Uskovo has 77 inhabitants, 59 male and 18 female, 33 householders and 20 other persons – in other words, co-owners. Only nine have families. When the people of Uskovo gathered around the jail, where we were taking tea, and when the women and children, being more curious, came up front, the crowd looked like a gypsy camp. Among the women there were actually several dark-skinned gypsies with sly, hypocritically sorrowful faces, and almost all the children were gypsies. Uskovo has a few convict gypsies whose bitter fate is shared by their families, who followed them voluntarily. I was slightly acquainted with two or three of the gypsy women. A week before my arrival at Uskovo I had seen them in Rykovskoye with rucksacks on their backs begging at people's windows.

The Uskovo inhabitants live very poorly. Only eleven desyatins of land are cultivated for grain and kitchen gardens – that is, almost one-fifth of a desyatin per homestead. All live at government expense, receiving prison rations which are not acquired cheaply because

68

they have to carry them on their backs over the roadless taiga from Derbinskoye.

After a rest, we set out at five o'clock in the afternoon on foot for Voskresenskoye. The distance is short, only six versts, but because of my inexperience in walking through the taiga I began to feel tired after the first verst. It was raining heavily. Immediately after leaving Uskovo we had to cross a stream about a sazhen wide on thin, crooked logs. My companions crossed safely, but I slipped and got my boot full of water. Before us lay a long, straight road cut through the forest for a projected highway. There was literally not one sazhen which you could walk without being thrown off balance or stumbling: hillocks, holes full of water, stiff tangles of bushes or roots treacherously concealed under the water, and against these you stumble as against a doorstep. The most unpleasant of all were the windfalls and the piles of logs cut down in order to carve out the road. You climb up one pile, sweat, and go on walking through the mud, and then you find another pile of logs and there is no way of bypassing it. So you start climbing again, while your companions shout that you are going the wrong way, it should be either left or right of the pile, etc. At the beginning I tried not to get my other boot full of water, but soon I gave up and resigned myself to it. I could hear the laboured breathing of the three settlers who were following behind, carrying our belongings. I was fatigued by the oppressive weather, shortness of breath and thirst. We walked without our service caps; it was easier.

The breathless general sat down on a thick log. We

sat down beside him. We gave a cigarette to each of the settlers, who dared not sit down.

'Well, it's hard going!'

'How many versts to Voskresenskoye?'

'Three more.'

A. M. Butakov walked the most briskly. He had formerly covered tremendous distances over the taiga and tundra, and a six-verst hike was nothing to him. He described his trip along the Poronay River and around Terpeniya Bay. The first day you are exhausted, all your strength gone, the second day your entire body aches but it is already becoming easier to walk; on the third and following days you feel you have sprouted wings, you are not walking but are being carried along by some unknown force, although your legs continue to get entangled in the merciless marsh grass and stuck in swamps.

Halfway it began to grow dark and soon we were surrounded by pitch darkness. I gave up hope that we would ever end our trip, and just groped ahead, splashing in water to my knees, and bumping into logs. Here and there the will-o'-the-wisps gleamed and flickered; entire pools and tremendous rotting trees were lit with phosphorescent colours and my boots were covered with moving sparks which shimmered like the glow-worms on a midsummer night.

But, thank God, at last a light shone in front of us, and was not phosphorescent, but a real light. Someone shouted at us, and we answered. The warden appeared with a lantern. Across pools brightly lit by his lantern, he came with large strides to lead us across the whole of Voskresenskoye, which was barely visible in the darkness, until at last we reached his quarters.

My companions had brought with them a change of clothing. When they reached the warden's quarters they hastened to change. But I had nothing with me, although I was literally soaked through. We drank some tea, talked a bit and went to sleep. There was only one bed in the warden's quarters, and this was taken by the general, while we ordinary mortals went to sleep on straw heaped on the floor.

Voskresenskoye is twice as large as Uskovo. Inhabitants, 183: 175 male and 8 female. There are 7 free families but not one legally married. There are few children in the settlement and only one little girl. It has 97 homesteaders and 77 co-owners.

iii. The Gilyak Tribe

Both of the northern districts, as the reader may readily see from my survey of the settlements, cover an area equal to a small Russian district. It is impossible to compute the area of both of them because there are no northern and southern boundaries. Between the administrative centres of both districts, the Alexandrovsk Post and Rykovskoye, there is a distance of 60 versts by the shorter route which crosses the Pilinga, while across the Arkovskaya valley it is 74 versts. In this kind of country these are large distances. Without considering Tangi and Vangi, even Palevo is considered a distant settlement. Meanwhile the newly founded settlements to the south of Palevo on the Poronaya tributaries raise the question of whether a new district will have to be established.

As an administrative unit, a Sakhalin district corresponds to a Russian district. According to the Siberian way of thinking, this term can only be applied to a postal distance which cannot be travelled in under a month, as for example the Anadyrsky district. To a Siberian official working alone in an area of 200 to 300 versts, the breaking up of Sakhalin into small districts would be a luxury. The Sakhalin population, however, lives under exceptional conditions and the administrative mechanism is far more complicated than in the Anadyrsky district. The need to break up the penal colonies into small administrative units has been shown by experience, and this has proved, in addition to other matters to be explained later, that the shorter the distances in the penal colony, the easier and more effective is the administration. Also, a breakup into smaller districts has the effect of enlarging the number of officials, and the result is an influx of new people who inevitably have a beneficial influence on the colony. And so with a quantitative increase of intelligent people on the staff, there occurs a significant increase in quality.

When I arrived in Sakhalin I heard a great deal of talk about a newly projected district. They described it as the Land of Canaan, because the plan called for a road which would cross the entire region southward along the Poronaya River. It was believed that the convicts at Dué and Voyevodsk would be transferred to the new district, and these horrifying places would become nothing more than a memory. Also, the mines would be taken away from the 'Sakhalin Company,' which had long since broken its contract, and then the mines would be worked by convicts and settlers as a collective enterprise.

Before completing my report on Northern Sakhalin, I feel I should discuss briefly a people who have lived here at different times and continue to live here outside the penal colony.

In the Duyka valley Polyakov found chipped obsidian knives, stone arrows, grinding stones, stone axes and other objects. He came to the conclusion that a people who did not use metal lived in the Dué valley in ancient times; they belonged to the Stone Age. Shards, the bones of dogs and bears, sinkers from large fishing nets, which were found in these formerly inhabited areas, indicate that they made pottery, hunted bear, went fishing and had hunting dogs. Clearly they derived flint from their neighbours on the mainland and on the neighbouring islands, because flint does not exist on Sakhalin. Probably the dogs played the same role in their migration as they do now: they are used for drawing sleighs. In the Tym valley Polyakov found the remnants of primitive structures and crude weapons. He concluded that in Northern Sakhalin 'it is possible for tribes to survive on a relatively low level of intellectual development; the people who lived here for centuries developed ways to protect themselves from cold, thirst and hunger. In all probability these ancient people lived in relatively small communities and were not a completely settled people.'

When sending Boshnyak to Sakhalin, Nevelskoy asked him to verify the rumour about people who had been left on Sakhalin by Lieutenant Khvostov and who had lived, according to the Gilyaks, on the Tym River.

Boshnyak was successful in discovering traces of these people. In one Tym River settlement the Gilyaks exchanged four pages torn from a prayerbook for three arshins of nankeen cloth, saying the prayerbook had been the property of Russians who had once lived there. On the title page, in barely legible script, were the words: 'We, Ivan, Danilo, Pyotr, Sergey and Vasily, were landed in the Aniva settlement of Tomari-Aniva by Khvostov on August 17, 1805. We moved to the Tym River in 1810 when the Japanese arrived in Tomari.' Later, exploring the area where the Russians had lived,

Boshnyak concluded that they had lived in three huts and cultivated gardens. The natives told him that the last of the Russians, Vasily, died recently, that they were fine people, that they went fishing and hunting with the natives and dressed native fashion except for cutting their hair. Elsewhere the natives informed him that two of the Russians had had children with native women. Today the Russians left by Khvostov on Northern Sakhalin have been forgotten and nothing is known of their children.

Boshnyak adds that as a result of his constant inquiries concerning any Russians settled on the island, he learned from natives in the Tangi settlement that some thirty-five or forty years ago there had been a shipwreck, the crew were saved and they built themselves first a house and later a boat. They made their way across La Pérouse to the Tatar Strait by boat and they were again shipwrecked near the village of Mgachi. This time only one man was saved. His name was Kemets. Not long afterward two Russians came from the Amur. Their names were Vasily and Nikita, and they joined Kemets and built themselves a house in Mgachi. They hunted game professionally and traded with the Manchurians and Japanese.

One Gilyak showed Boshnyak a mirror supposedly given to his father by Kemets. The Gilyak would not sell the mirror at any price, saying that he was keeping it as a precious memento of his father's friend. Vasily and Nikita were terrified of the Tsar, and it is obvious that they had escaped from his prisons. All three died on Sakhalin.

The Japanese Mamia-Rinzo learned in 1808 on Sakhalin that Russian boats often appeared on the western side of the island, and the piracy practised by the Russians eventually forced the natives to expel one group of Russians and to massacre another. Mamia-Rinzo names these Russians as Kamutsi, Simena, Momu and Vasire. 'The last three,' says Shrenk, 'are easily recognizable as the Russian names Semyon, Foma and Vasily. Kamutsi is quite similar to Kemets,' in his opinion.

This short history of eight Sakhalin Robinson Crusoes exhausts all the data concerning the free colonization of Northern Sakhalin. If the extraordinary fate of five of Khvostov's sailors, Kemets and the two refugees from prison resembles an attempt at free colonization, this attempt must be regarded as insignificant and completely unsuccessful. The really important fact is that they all lived on Sakhalin for a long time, and to the end of their lives not one of them engaged in agriculture. They lived by fishing and hunting.

To round out the picture I must mention the local indigenous population – the Gilyaks. They live on the western and eastern banks of Northern Sakhalin and along the rivers, especially the Tym.

The villages are old; their names, mentioned in the writings of old authors, have come down without change. However, their way of life cannot be called completely settled, because a Gilyak feels no ties toward his birthplace or to any particular place. They often leave their yurts to practise their trades, and to wander over Northern Sakhalin with their families and dogs. But as to their wanderings, even when they are forced to take long journeys to the mainland, they remain faithful to the island, and the Sakhalin Gilyak differs in language and customs from the Gilyak living on the mainland no less than the Ukrainian differs from the Muscovite.

In view of this, it seems to me that it would not be very difficult to count the number of Sakhalin Gilyaks without confusing them with those who come for trading purposes from the Tatar shore. There would be no harm in taking a census of them every five to ten years; otherwise the important question of the influence of the penal colony on their numbers will long remain open and will be solved in a quite arbitrary fashion.

According to data gathered by Boshnyak in 1856, there were 3,270 Gilyaks on Sakhalin. Fifteen years later Mitsul found only 1,500, and the latest data which I obtained from the prison copy of *Statistical Records of Foreigners*, 1889, showed there were only 320 Gilyaks in both regions. If these figures hold true, not one Gilyak will remain in ten or fifteen years' time. I cannot judge the correctness of the figures given by Boshnyak and Mitsul, but the official figure of 320 can have no significance whatsoever. There are several reasons for this. Statistics on foreigners are calculated by clerks who have neither the educational background nor the practical knowledge to do it, and they are given no instructions. When they gather information at the Gilyak settlements, they naturally conduct themselves in an overbearing manner. They are rude and disagreeable, in contrast to the polite Gilyaks, who do not permit an arrogant and domineering attitude toward people. Because they are averse to any kind of census or registration, considerable skill is needed in handling them. Also, the data is gathered by the administration without any definite plan, only in passing, and the investigator uses no ethnographic map but works in his own arbitrary fashion. The data on the Alexandrovsk district includes only those Gilyaks who live south of the Vangi settlement, while in the Tymovsky district they counted only those they found near the Rykovskoye settlement. Actually they do not live in this settlement, but pass through it on their way to other places.

Undoubtedly the number of Sakhalin Gilyaks is constantly decreasing, and this judgment can be made simply by eye-count. How large is this decrease? Why is it taking place? Is it because Gilyaks are becoming extinct, or because they are moving to the mainland or farther north on the island? Due to the lack of actual statistics (and our figures on the destructive influence of Russian colonization can be based only on analogies) it is quite possible that up to the present day Russian influence has been insignificant, almost zero, since the Sakhalin Gilyaks live by preference along the Tym and the eastern shores of the island, which the Russian settlements have not yet reached.

The Gilyaks are neither Mongols nor Tungus, but belong to some unknown race which may once have been powerful and ruled all of Asia. Now, living out their last centuries on a small patch of land, they are only a small remnant. Yet they are a wonderful and cheerful people. Because of their unusual sociability and mobility, the Gilyaks long ago succeeded in having relations with all the neighbouring peoples, and so it is almost impossible to find a pure-blooded Gilyak without Mongol, Tungus or Ainu elements.

A Gilyak's face is round, flat, moonlike, of yellowish cast, with prominent cheekbones, dirty, with slanting eyes and a barely visible beard. His hair is smooth, black, wiry, gathered into a braid at the nape of the neck. His facial expression is not savage; it is always intelligent, gentle, naïvely attentive; he is either blissfully smiling or thoughtfully mournful, like a widow. When he stands in profile with his sparse beard and braid, with a soft, womanish expression, he could be a model for a picture of Kuteykin, and it becomes almost understandable why some travellers regard the Gilyaks as belonging to the Caucasian race.

Anyone who wants to become thoroughly acquainted with the Gilyaks should consult an ethnographic specialist, L. I. Shrenk. I will limit myself to discussing some of the characteristics of local natural conditions, which may be useful as direct or indirect guidance for new colonists.

The Gilyak has a strong, stocky build, and he is of medium or short stature. Height would be of no advantage to him in the taiga. His bones are thick and distinguished by the strong development of his limbs from rowing and tramping over the hills. This exercise strengthens the muscles, and they indicate powerful musculature and a perpetual, intense struggle against nature. His body is lean, without fat. There are no stout or corpulent Gilyaks. All his fat is used for the warmth which a man on Sakhalin must generate in his body in order to compensate for the heat loss caused by the low temperature and the excessive humidity. It is understandable that a Gilyak should require a good deal of fat in his diet. He eats fatty seal meat, salmon, sturgeon and whale fat. He also eats rare meat in large quantities in raw, dry and frozen form, and because he eats coarse food his chewing muscles are unusually well-developed and all his teeth are badly worn. His food consists exclusively of meat but on rare occasions, at home or while carousing, they add Manchurian garlic or berries to their menus. According to Nevelskoy, the Gilyaks consider agriculture a grievous transgression; whoever ploughs the land or plants anything will soon die. But they eat the bread which the Russians introduced to them with relish, as a delicacy, and it is not unusual to see a Gilyak in Alexandrovsk or Rykovskoye carrying a loaf of bread under his arm.

The Gilyak's clothing has been adapted to the cold, damp and rapidly changing climate. In the summer he wears a shirt of blue nankeen or daba cloth with trousers of the same material. Over his back, as insurance against changing weather, he wears either a coat or a jacket made of seal or dog fur. He puts on fur boots. In winter he wears fur trousers. All this warm clothing is cut and sewn so as not to impede his deft and quick movements while hunting or while riding with his dogs. Sometimes, in order to be in fashion, he wears convict overalls. Eighty-five years ago Krusenstern saw a Gilyak dressed in a magnificent silk costume 'with many flowers woven into it.' Today you will not find such a peacock on Sakhalin if you search with a lamp.

As to Gilyak yurts, these again answer the demands of a damp and cold climate. There are both summer and winter yurts. The first are built on stilts, the second are dug-outs with timber walls having the form of a truncated pyramid. The wood outside is covered with sod. These yurts are made of cheap material which is always at hand, and when the necessity arises they have no regret at leaving them. They are warm and dry, and are certainly far superior to the damp and cold huts made of bark in which our convicts live when they are working on roads or in the fields. These summer yurts should positively be recommended for gardeners, charcoal makers, fishermen and all convicts and settlers who work outside the prison and not in their homes.

Gilyaks never wash, with the result that even ethnographers find it difficult to ascertain the colour of their skins. They never wash their underclothing, and their furs and boots look exactly as if they had just been stripped off a dead dog. The Gilyaks themselves exude a heavy, sharp odour and the close proximity of their dwellings is indicated by the foul, almost unbearable odour of drying fish and rotting fish wastes. Usually near every yurt there is a drying contrivance which is filled to the top with flattened fish, which from afar, especially in the sunshine, looks like strings of coral. Krusenstern found huge masses of tiny maggots an inch thick on the ground surrounding these fish driers.

In the winter the yurts are full of pungent smoke issuing from the hearth. In addition, the Gilyak men, their wives and even the children smoke tobacco.

Nothing is known of the diseases and mortality of the Gilyaks, but it may be supposed that the unhealthy, unhygienic circumstances are detrimental to their health. This may be the cause of their short stature, bloated faces, the sluggishness and laziness of their movements; and this is perhaps why the Gilyaks always have weak resistance to epidemics. The devastation on Sakhalin caused by smallpox is well known.

Krusenstern found twenty-seven houses on Sakhalin's northernmost point, between the Elizaveta and Maria capes. In 1860, P. P. Glen, a participant in a wonderful Siberian expedition, found only traces of the settlement, while in other parts of the island, he tells us, he found evidence that there was once a considerable population. The Gilyaks told him that during the past ten years – i.e., after 1850 – the population had been radically reduced by smallpox. It is certain that the terrible smallpox epidemics which devastated Kamchatka and the Kurile Islands did not bypass Sakhalin. Naturally this was not due to the virulence of the smallpox itself but to the Gilyaks' poor ability to resist it. If typhus or diphtheria are brought into the penal colony and reach the Gilyak yurts, the same effect will be achieved as by the smallpox. I did not hear of any epidemics on Sakhalin; it seems there were none for the past twenty years with the exception of an epidemic of conjunctivitis, which can be observed even now.

General Kononovich gave permission to the regional

hospitals to accept non-Russian patients at government expense (Order No. 335, 1890). We have no exact observations of Gilyak diseases, but some inferences can be drawn as to the causes of their diseases: dirtiness, excessive use of alcohol, intercourse with Chinese and Japanese, constant closeness to dogs, traumas, etc., etc.

There is no doubt they have frequent illnesses and require medical assistance, and if circumstances permit them to take advantage of the new order granting them admission to the hospitals, the local doctors will have the opportunity of studying them more closely. Medicine cannot arrest their yearly mortality, but perhaps the doctors may discover the circumstances under which our interference with the lives of these people will be least harmful.

The character of the Gilyaks is described in different ways by different authors, but all agree that they are not aggressive, dislike brawls and quarrels, and live peacefully with their neighbours. When strangers appear, they are always suspicious and apprehensive; nevertheless, they greet them courteously, without any protest, and sometimes they will lie, describing Sakhalin in the worst possible light, hoping in this way to discourage strangers from the island. They embraced Krusenstern's companions, and when L. I. Shrenk became ill, the news quickly spread among the Gilyaks and evoked the deepest sympathy.

They lie only when they are trading or when speaking to someone they look upon with suspicion, who is therefore in their eyes dangerous, but before telling a lie they always look at one another – a distinctive childish trait. All other lying and boasting in daily life, outside of trading, is repugnant to them.

The following incident occurred early one evening. Two Gilyaks, one with a beard and the other with a swollen feminine face, lay on the grass in front of a settler's hut. I was passing by. They called out to me and started begging me to enter the hut and bring out their outer clothing, which they had left at the settler's that morning. They themselves did not dare to go in. I told them I had no right to go into someone's hut in the absence of the owner. They grew silent.

'You are a politician?' asked the feminine-looking Gilyak in bad Russian.

'No.'

'That means you are a *pishi–pishi*?' [*pisar* means clerk] he said, seeing some paper in my hands.

'Yes, I write.'

'How much salary do you get?'

I was earning about 300 roubles a month. I told them the figure. You should have seen the disagreeable and even painful expressions which my answer produced. Both Gilyaks suddenly grabbed their stomachs, and throwing themselves on the ground, they began rolling around exactly as though they had severe stomach cramps. Their faces expressed despair.

'How can you talk that way?' they said. 'Why did you say such an awful thing? That's terrible! You shouldn't do that!'

'What did I say that was bad?' I asked.

'Butakov, the regional superintendent, well, he's a big man, gets 200, while you are not even an official –

a clerk – amounts to nothing, and you get 300! You spoke untruth! You shouldn't do that!'

I tried to explain that a regional superintendent remains in one place and therefore only gets 200 roubles. Although I am just a '*pishi-pishi*,' I have come a long way – 10,000 versts away. My expenses are greater than Butakov's and therefore I need more money. This calmed the Gilyaks. They exchanged glances, spoke together in Gilyak, and stopped suffering. Their faces showed that they finally believed me.

'It's true, it's true!' said the bearded Gilyak briskly. 'That's fine. You may leave now!'

'It's true,' nodded the other. 'You may go!'

When a Gilyak accepts an obligation, he fulfills it properly. There has never been a single case of a Gilyak dumping mail along the road or embezzling the property of others. Polyakov, who had dealings with Gilyak boatmen, wrote that they were most punctilious in fulfilling an obligation, and this is characteristic of them today when we find them unloading government freight for the prisons.

They are clever, intelligent, cheerful, brash, and are never shy in the society of strong and rich men. They do not accept authority, and they do not even understand the meaning of 'older' and 'younger.' In *The History of Siberia*, by I. Fisher, we read that the renowned Polyakov visited the Gilyaks, who were then 'under no foreign domination.' They have a word, *dzhanchin*, which denotes 'your excellency,' and they use it equally to a general or to a rich trader who has a great deal of nankeen and tobacco. Seeing Nevelskoy's picture of the Tsar, they said he must be a strong man who distributes much nankeen and tobacco.

The commandant of the island possesses vast and terrifying powers. Nevertheless, when I was riding with him from Verkhny Armudan to Arkovo, a Gilyak had no compunction about shouting at us imperiously: 'Stop!' Then he asked if we had seen his white dog along the road.

As it is often said and written, Gilyaks have no respect for family seniority. A father does not believe he is senior to his son, and a son does not respect his father, but lives as he pleases. An old mother has no more authority in the yurt than a teenage daughter. Boshnyak wrote that he often saw a son beat his mother and chase her out of the house and no one dared say a word against him. The male members of a family are equal to one another. If you treat Gilyaks to vodka, it must also be served to the very youngest males.

The females are equally without rights, whether it is a grandmother, mother or breast-fed baby girl. They are treated as domestic animals, as chattels, which can be thrown out, sold or kicked like a dog. The Gilyaks pet their dogs, but women – never. Marriage is considered nonsense – much less important, for example, than a drinking bout. It is not accompanied by any religious or superstitious rites. A Gilyak exchanges a spear, a boat or a dog for a young girl, drives her to his yurt and lies down with her on a bearskin – and that is all there is to it. Polygamy is permitted but is not widespread, although there are obviously more women than men. Contempt for women as for a lower creature or possession has come to such a pass that the Gilyak does not consider slavery, in the exact and coarse meaning of the word, as reprehensible. As Shrenk witnessed, the Gilyaks often bring Ainu women home with them as slaves. Plainly a woman is an object of barter, like tobacco or daba cloth. Strindberg, that famous misogynist, who thought women should be slaves of men's desires, follows the Gilyak pattern. If he happened to visit Northern Sakhalin, they would embrace him warmly.

General Kononovich told me he wants to Russify the Sakhalin Gilyaks. I don't know why this is necessary. Furthermore, Russification had already begun long before the general's arrival. It began when some prison wardens, receiving very small salaries, began acquiring expensive fox and sable cloaks at the same time that Russian water jars appeared in Gilyak yurts.

As time passed, the Gilyaks were hired to help in tracking down prisoners who escaped from the prison. There was a reward for capturing them, dead or alive. General Kononovich ordered Gilyaks to be hired as jailers. One of his orders says this is being done because of the dire need for people who are well acquainted with the countryside, and to ease relations between the local authorities and the non-Russians. He told me personally that his new ruling is also aimed at their Russification.

The first ones approved as jailers were the Gilyaks Vaska, Ibalka, Orkun and Pavlinka (Order No. 308, 1889). Later, Ibalka and Orkun were discharged 'for continuous failure to appear at the administrative office

to receive their orders,' and they then approved Sof-
ronka (Order No. 426, 1889). I saw these jailers; they
wore tin badges and revolvers. The most popular and
the one who is seen most often is the Gilyak Vaska, a
shrewd, sly drunkard. One day I went to the shop
supported by the colonization fund and met a large
group of the intelligentsia. Someone, pointing at a shelf
full of bottles, said that if you drank them all down you
would really get drunk, and Vaska smirked fawningly,
glowing with the wild joy of a tippler. Just before my
arrival a Gilyak jailer on duty killed a convict and the
local sages were concerned with only one question:
whether he was shot in the chest or in the back – that
is, whether to arrest the Gilyak or not.

That their proximity to the prison will not Russify
but eventually alienate the Gilyaks does not have to be
proved. They are far from understanding our require-
ments, and there is scarcely any opportunity to explain
to them that convicts are caught, deprived of their
freedom, wounded and killed not because of caprice,
but in the interests of justice. They regard this as
coercion, a display of bestiality, and probably consider
themselves as hired killers.

If it is absolutely necessary to Russify them and if it
cannot be avoided, I believe that when we choose our
methods, our primary concern should not be our own
needs, but theirs. The order permitting them to
become patients in our hospitals, the distribution of
aid in the form of flour and groats, as was done in 1886
when the Gilyaks were starving, and the order not to
confiscate their property for debt, and the remission

of their debts (Order No. 204, 1890), and all similar measures will probably achieve this aim more quickly than tin badges and revolvers.

In addition to the Gilyaks, there are a small number of Oroki, or Orochi, of the Tungus tribe living in Northern Sakhalin. Since they are barely heard of in the colony and since no Russian settlements exist in this area, I merely mention them here.

iv. *The Morality of Sakhalin*

Some convicts bear their punishment with fortitude, readily admit their guilt, and when you ask them why they came to Sakhalin, they usually answer, 'They do not send anyone here for their good deeds.' Others astonish you with their cowardice and the melancholy face they show to the world. They grumble, they weep, are driven to despair, and swear they are innocent. One considers his punishment a blessing because, he says, only in penal servitude did he find God. Another attempts to escape at the first opportunity, and when they catch up with him, he turns on his captors and clubs them. Accidental transgressors, 'unfortunates,' and those innocently sentenced live under one roof with inveterate and incorrigible criminals and outcasts.

When the general question of morality is discussed, we must admit that the exile population produces an extremely mixed and confusing impression, and with the existing means of research it is scarcely possible to form any serious generalizations. The morality of a population is usually judged by statistics of crimes, but even this normal and simple method is useless in a penal colony. The strictly nominal infractions of the law, the self-imposed rules and the transgressions of the convict population living under abnormal and exceptional conditions – all these things which we consider petty violations are regarded as serious crimes in Sakhalin, and conversely a large number of serious crimes committed here are not regarded as crimes at all, because they are considered perfectly normal and inevitable phenomena in the atmosphere of the prison.

The vices and perversions which may be observed among the exiles are those which are peculiar to enslaved, subjected, hungry and frightened people. Lying, cunning, cowardice, meanness, informing, robbery, every kind of secret vice – such is the arsenal which these slavelike people, or at least the majority of them, employ against the officials and guards they despise, fear and regard as their enemies. The exile resorts to deceit in order to evade hard labour or corporal punishment and to secure a piece of bread, a pinch of tea, salt or tobacco, because experience has proved to him that deceit is the best and most dependable strategy in the struggle for existence. Thievery is common and is regarded as a legitimate business.

The prisoners grab up everything that is not well hidden with the tenacity and avarice of hungry locusts, and they give preference to edibles and clothing. They steal from each other in the prison; they steal from settlers, and at their work, and when loading ships. The virtuosity of their dexterous thieving may be judged by the frequency with which they practice their art. One day they stole a live ram and a whole tub of sour dough from a ship in Dué. The barge had not yet left the ship, but the loot could not be found. On another occasion they robbed the commander of a ship, unscrewing the lamps and the ship's compass. On still another occasion they entered the cabin of a foreign ship and stole the silverware. During the unloading of cargo whole bales and barrels vanish.

A convict takes his recreation secretly and furtively. In order to obtain a glass of vodka, which under ordinary circumstances costs only five kopecks, he must surreptitiously approach a smuggler and if he has no money he must pay in bread or clothing. His sole mental diversion – card-playing – can only be enjoyed at night in the light of candle stubs or in the taiga. All secret amusements, when repeated, slowly develop into passions. The extreme imitativeness among the convicts causes one prisoner to infect another and finally such seeming inanities as contraband vodka and card-playing lead to unbelievable lawlessness. As I have already said, kulaks among the convicts often amass fortunes. This means that alongside convicts who possess 30,000 to 50,000 roubles, you find people who systematically squander their food and clothing.

Card-playing has infected all the prisons like an epidemic. The prisons are large gambling houses, while the settlements and posts are their branches. Gambling is exceptionally widespread. They say that during a chance search the organizers of the local card games were found to be in possession of hundreds and thousands of roubles, and they are in direct communication with the Siberian prisons, notably the prison at Irkutsk, where, so the prisoners say, they play 'real' cards.

There are several gambling houses in Alexandrovsk. There was even a scandal in a gambling house on Second Kirpichnaya Street, which is characteristic of similar haunts. A guard lost everything and shot himself. Playing faro dulls the brain and acts like a narcotic. The convict loses his food and clothing, feels neither hunger nor cold, and suffers no pain when he is beaten. And how strange it is that even when they are loading a ship, and the coal barge is bumping broadside against the ship, and the waves are smashing against it and they are growing green with seasickness, even then they play cards on the barge and casual everyday expressions are mingled with words which arise purely from card-playing: 'Push off!' 'Two on the side!' 'I've got it!'

Furthermore, the subservient status of the woman, her poverty and degradation, are conducive to the development of prostitution. When I asked in Alexandrovsk if there were any prostitutes there, they answered, 'As many as you want.' Because of the tremendous demand, neither old age, nor ugliness, nor even tertiary syphilis is an impediment. Even extreme youth is no hindrance. I met a sixteen-year-old girl on a street in Alexandrovsk, and they say she has been engaged in prostitution since she was nine years old. The girl has a mother, but a family background on Sakhalin does not always save a young girl from disaster. They talk about a gypsy who sells his daughters and even haggles over them. One free woman in Alexandrovsk has an 'establishment,' in which only her own daughters operate. In Alexandrovsk corruption is generally of an urban character. There are 'family baths' run by a Jew, and the names of the professional panderers are known.

According to government data, incorrigible criminals, those who have been resentenced by the district court, comprise 8 per cent of the convicts as of

January 1, 1890. Among the incorrigibles were some who have been sentenced three, four, five and even six times. There are 175 persons who through their incorrigibility have spent twenty to fifty years in penal servitude – i.e., 3 percent of the total. But these are exaggerated figures for incorrigibles, since the majority of them were shown to have been resentenced for attempts to escape. And these figures are inaccurate with regard to attempted escapes, because those who have been caught are not always brought to trial but are most frequently punished in the usual fashion. The extent to which the exile population is delinquent or, in other words, criminally inclined is presently unknown.

True, they do try people here for crimes, but many cases are dismissed because the culprits cannot be found, many are returned for additional information or clarification of jurisdiction, or the trial remains at a standstill because the necessary information has not been received from the various Siberian offices. Finally, after a great deal of red tape, the documents go into the archives upon the death of the accused, or if nothing more is heard of him after his escape. Credence is attached to evidence by young people who have received no education, while the Khabarovsk court tries people from Sakhalin *in absentia*, basing its verdict only on documents.

During 1889, 243 convicts were under juridical investigation or on trial, that is, one defendant for every 25 convicts. There were 69 settlers under investigation and on trial, i.e., one in 55. Only 4 peasants were under investigation, i.e., one in 115. From these ratios it is evident that with the easing of his lot and with the transition of the convict to a status giving him more freedom, the chances of being brought to trial are decreased by half each time. All these figures pertain to persons on trial and under investigation, but do not necessarily represent crimes committed in 1889, because the files dealing with these crimes refer to trials begun many years ago and not yet completed. These figures give the reader some idea of the tremendous number of people on Sakhalin who languish year after year in the courts and under investigation, because their cases have been drawn out over a period of years. The reader can well imagine how destructively this system reacts on the economy and on the spirit of the people.

Investigation is usually entrusted to the prison warden's assistant or to the secretary of the police department. According to the island commandant, 'investigations are begun on insufficient information, they are conducted sluggishly and clumsily, and the prisoners are detained without any reason.' A suspect or an accused person is arrested and put in a cell. When a settler was killed at Goly Mys, four men were suspected and arrested. They were placed in dark cold cells. In a few days three were released, and only one was detained. He was put in chains and orders were issued to give him hot food only every third day. Then, by order of the warden, he was given 100 lashes. A hungry, frightened man, he was kept in a dark cell until he confessed. The free woman Garanina was detained in the prison at the same time on suspicion of having murdered her husband. She was also placed in a dark cell and received hot food every third day. When one official questioned her in my presence, she said that she had been ill for a long time and that for some reason they would not permit a doctor to see her. When the official asked the guard in charge of the cells why they had not troubled to get a doctor for her, he answered, 'I went to the honourable warden, but he only said, "Let her croak." '

This incapacity to differentiate imprisonment before trial from punitive imprisonment (and this in a dark cell of a convict prison), the incapacity to differentiate between free people and convicts amazed me especially because the local district commander is a law-school graduate and the prison warden was at one time a member of the Petersburg Police Department.

I visited the cells a second time early in the morning in the company of the island commandant. Four convicts suspected of murder were released from their cells; they were shivering with cold. Garanina, in stockings and without shoes, was shivering and blinking in the light. The commandant ordered her transferred to a room with good light. I saw a Georgian flitting like a shadow around the entrance to the cells. He has been held for five months in the dark hallway on suspicion of poisoning and is awaiting investigation. The assistant prosecutor does not live on Sakhalin and there is nobody to supervise an investigation. The direction and speed of an investigation are totally dependent on various circumstances which have no reference to the case itself. I read in one report that the murder of a certain Yakovleva was committed 'with the intent of robbery with a preliminary attempt at rape, which is evidenced by the rumpled bedding and fresh scratches and impressions of heel spikes on the backboard of the bed.' Such a consideration predetermines the outcome of the trial; an autopsy is not considered necessary in such cases. In 1888 an escaped convict murdered Private Khromatykh and the autopsy was only conducted in 1889 on the demand of the prosecutor when the investigation had been completed and the case brought to trial.

Article 469 of the *Code* permits the local administration to specify and carry out punishment without any formal police investigation for such crimes and offences by criminals for which punishment is due according to the general criminal laws, not excluding the loss of all personal rights and privileges in imprisonment. Generally the petty cases on Sakhalin are judged by a formal police court which is under the authority of the police department. Notwithstanding the broad scope of this local court, which has jurisdiction over all petty crimes as well as over a multitude of cases which are only nominally regarded as petty, the local community does not enjoy justice and lacks a court of law. Where an official has the right, according to law, to flog and incarcerate people without trial and without investigation and even to send them to hard labour in the mines, the existence of a court of law has merely formal significance.

Punishment for serious crimes is decided by the Primorskaya district court, which settles cases only on documentary evidence without questioning the defendants or witnesses. The decision of a district court is always presented for approval to the island commandant, who, if he disagrees with the verdict, settles the case on his own authority. If the sentence is changed, the fact is reported to the ruling senate. If the administration considers a crime as being more serious than it appears to be on the official record, and if it regards the punishment as insufficient according to the *Code on Convicts*, then it petitions for arraignment of the defendant before a court-martial.

The punishment usually inflicted upon convicts and settlers is distinguished by extraordinary severity. Our *Code on Convicts* is at odds with the spirit of the times and of the laws, and this is especially evident in the sections concerning punishment. Punishments which humiliate the offender, embitter him and contribute to his moral degradation, those punishments which have long since been regarded as intolerable among free men, are still being used here against settlers and convicts. It is as though exiles were less subject to the dangers of becoming bitter and callous, and losing their human dignity. Birch rods, whips, chains, iron balls, punishments which shame the victim and cause pain and torment to his body, are used extensively. Floggings with birch rods and whips are habitual for all kinds of transgressions, whether small or large. It is the indispensable mainstay of all punishment, sometimes supplementing other forms of chastisement, or used alone.

The most frequently used punishment is flogging

with birch rods. As shown in the official report, this punishment was imposed on 282 convicts and settlers in Alexandrovsk in 1889 by orders of the administration: corporal punishment, i.e., with birch rods, was inflicted on 265, while 17 were punished in other ways. The administration used birch rods in 94 out of 100 cases. In fact, the number of criminals suffering corporal punishment is far from being accurately recorded in the reports. The reports of the Tymovsky district for 1889 show that only 57 convicts were beaten with birch rods and only 3 are recorded in Korsakov; although the truth is that they flog several people every day in both districts, and sometimes there are 10 a day in Korsakov.

All sort of transgressions may result in a man's getting 30 to 100 strokes with birch rods: nonperformance of the daily work quota (for example, if the shoemaker did not sew his required three pairs of shoes), drunkenness, vulgarity, insubordination . . . If 20 to 30 men fail to complete their work quota, all 20 to 30 are beaten. One official told me:

The prisoners, especially those in irons, like to present absurd petitions. When I was appointed here, I toured the prison and received 50 petitions. I accepted them, and then announced that those whose petitions do not deserve attention would be punished. Only 2 of the petitions proved to be worthwhile, the remainder were nonsense. I ordered 48 men to be flogged. The next time 25 were flogged, and later fewer and fewer, and now they no longer send me petitions. I cured them of the habit.

In the South, as a result of a convict's denunciation, a search was made of another convict's possessions and a diary was found which was presumed to contain drafts of correspondence carried on with friends at home. They gave him 50 strokes with birch rods and kept him 15 days in a dark cell on bread and water. With the knowledge of the district commander, the inspector in Lyutoga gave corporal punishment to nearly everyone. Here is how the island commandant describes it:

The commander of the Korsakov district informed me about the extremely serious instances of excessive authority used by X., who ordered some settlers to receive corporal punishment far beyond the limits set by the law. This instance, shocking in itself, is even more shocking when the circumstances which provoked the punishment are analysed. There had been a quite commonplace and futile brawl between exiled settlers; and it made no difference to him whether he punished the innocent or the guilty, or pregnant women. [Order No. 258, 1888.]

Usually an offender receives 30 to 100 strokes with birch rods. This depends on who gave the order to punish him, the district commander or the warden. The former has the right to order up to 100, the latter up to 30. One warden always gave 30. Once when he was required to take the place of the district commander, he immediately raised his customary allotment to 100, as though this hundred strokes with birch rods was an indispensable mark of his new authority. He did not change the number until the district commander returned, and then in the same conscientious manner he resumed the old figure of 30. Because of its very frequent application, flogging with birch rods has become debased. It no longer causes abhorrence or fear among many prisoners. They tell me that there are quite a number of prisoners who do not feel any pain when they are being flogged with birch rods.

Lashes are used far less frequently and only after a sentence passed by the district courts. From a report of the director of the medical department it appears that in 1889, 'in order to determine the ability to endure corporal punishment ordered by the courts,' 67 prisoners were examined by the doctors. Of all the punishments exacted on Sakhalin this punitive measure is the most abominable in its cruelty and abhorrent circumstances, and the jurists of European Russia who sentence vagrants and incorrigible criminals to be flogged would have renounced this mode of punishment long ago had it been carried out in their presence. But these floggings are prevented from being a scandalous and outrageously sensational spectacle by Article 478 of the *Code*, which specifies that the sentences of the Russian and Siberian courts must be executed in the place where the prisoner is confined.

I saw how they flog prisoners in Dué. Vagrant Prokhorov, whose real name was Mylnikov, a man thirty-five to forty years of age, escaped from the Voyevodsk prison, and after building a small raft, he took off for the mainland. On shore they noticed him in time, and sent a cutter to intercept him. The investigation of his escape began. They took a look at the official records and then they made a discovery: this Prokhorov was actually Mylnikov, who had been sentenced last year by the Khabarovsk district court to 90 lashes and the ball and chain for murdering a Cossack and his two grandchildren. Owing to an oversight the sentence had not yet been carried out. If Prokhorov had not taken it into his head to escape, they might never have noticed their error and he would have been spared a flogging and being chained to an iron ball. Now, however, the execution of the sentence was inevitable.

On the morning of the appointed day, August 13, the warden, the physician and I leisurely approached the prison office. Prokhorov, whose presence in the office had been ordered the previous evening, was sitting on the porch with a guard. He did not know what awaited him. Seeing us, he got up. He may have understood then what was going to happen, because he blanched.

'Into the office!' the warden ordered.

We entered the office. They led Prokhorov in. The doctor, a young German, ordered him to strip and listened to his heart to ascertain how many lashes the prisoner could endure. He decides this question in a minute and then in a businesslike fashion sits down to write his examination report.

'Oh, the poor fellow!' he says sorrowfully in a thick German accent, dipping the pen into the ink. 'The chains must weigh upon you! Plead with the honourable warden and he will order them removed.'

Prokhorov remains silent. His lips are pale and trembling.

'Your hope is in vain,' the doctor continues. 'You all have vain hopes. Such suspicious people in Russia! Oh, poor fellow, poor fellow!'

The report is ready. They include it with the documents on the investigation of the escape. Then follows utter silence. The clerk writes, the doctor and the warden write. Prokhorov does not yet know exactly why he was brought here. Is it only because he escaped, or because of the escape and the old question as well? The uncertainty depresses him.

'What did you dream of last night?' the warden asks finally.

'I forgot, your worship.'

'Now listen,' says the warden, glancing at the official documents. 'On such and such a date you were sentenced to 90 lashes by the Khabarovsk district court for murdering a Cossack . . . And today is the day you are to get them.'

Then he smacks the prisoner on his forehead with the flat of his hand and admonishes him:

'Why did all this have to happen? It's because your head needs to be smarter than it is. You all try to escape and think you will be better off, but it turns out worse.'

We all enter the 'guardhouse,' which is a grey barracks-type building. The military medical assistant, who stands at the door, says in a wheedling voice as though asking a favour:

'Your worship, please let me see how they punish a prisoner.'

In the middle of the guardroom there is a sloping bench with apertures for binding the hands and feet.

The executioner is a tall, solid man, built like an acrobat. His name is Tolstykh. He wears no coat, and his waistcoat is unbuttoned. He nods at Prokhorov, who silently lies down. Tolstykh, taking his time, silently pulls down the prisoner's trousers to the knees and slowly ties his hands and feet to the bench. The warden looks callously out the window, the doctor strolls around the room. He is carrying a vial of medicinal drops in his hands.

'Would you like a glass of water?' he asks.

'For God's sake, yes, your worship.'

At last Prokhorov is tied up. The executioner picks up the lash with three leather thongs and slowly straightens it.

'Brace yourself!' he says softly, and without any excessive motion, as though measuring himself to the task, he applies the first stroke.

'One-ne,' says the warden in his chanting voice of a cantor.

For a moment Prokhorov is silent and his facial expression does not change, but then a spasm of pain runs along his body, and there follows not a scream but a piercing shriek.

'Two,' shouts the warden.

The executioner stands to one side and strikes in such a way that the lash falls across the body. After every five strokes he goes to the other side and the prisoner is permitted a half-minute rest. Prokhorov's hair is matted to his forehead, his neck is swollen. After the first five or ten strokes his body, covered by scars from previous beatings, turns blue and purple, and his skin bursts at each stroke.

Through the shrieks and cries there can be heard the words: 'Your worship! Your worship! Mercy, your worship!'

And later, after 20 or 30 strokes, he complains like a drunken man or like someone in delirium:

'Poor me, poor me, you are murdering me . . . Why are you punishing me?'

Then follows a peculiar stretching of the neck, the noise of vomiting. Prokhorov says nothing; only shrieks and wheezes. A whole eternity seems to have passed since the beginning of the punishment. The warden cries, 'Forty-two! Forty-three!' It is a long way to 90.

I go outside. The street is quite silent, and it seems to me that the heartrending sounds from the guard-house can be heard all over Dué. A convict wearing the clothing of a free man passes by and throws a fleeting glance in the direction of the guardhouse, terror written on his face and on his way of walking. I return to the guardhouse, and then go out again, and still the warden keeps counting.

Finally, 90! Prokhorov's hands and feet are quickly released and he is lifted up. The flesh where he was beaten is black and blue with bruises and it is bleeding. His teeth are chattering, his face yellow and damp, and his eyes are wandering. When they give him the medicinal drops in a glass of water, he convulsively bites the glass . . . They soak his head with water and lead him off to the infirmary.

'That was for the murder. He'll get another one for escaping,' I was told as we went home.

'I love to see how they execute punishment!' the military medical assistant exclaims joyfully, extremely pleased with himself because he was satiated with the abominable spectacle. 'I love it! They are such scum, such scoundrels. They should be hanged!'

Not only do the prisoners become hardened and brutalized from corporal punishment, but those who inflict the punishment become hardened, and so do the spectators. Educated people are no exception. At any rate, I observed that officials with university training reacted in exactly the same way as the military medical assistants or those who had completed a course in a military school or an ecclesiastical seminary. Others become so accustomed to birch rods and lashes and so brutalized that in the end they come to enjoy the floggings.

They tell a story about one prison warden who whistled when a flogging was being administered in his presence. Another warden, an old man, spoke to the prisoner with happy malice, saying, 'God be with you! Why are you screaming? It's nothing, nothing at all! Brace yourself! Beat him, beat him! Scourge him!' A third warden ordered the prisoner to be tied to the bench by his neck so that he would choke. He administered five or ten strokes and then went out somewhere for an hour or two. Then he came back and gave him the rest.

The courts-martial are composed of local officers

appointed by the island commandant. The documents on the case and the court's verdict are sent to the Governor-General for confirmation. In the old days prisoners languished in their cells for two and three years while awaiting confirmation of the sentence; now their fate is decided by telegraph. The usual sentence of the courts-martial is death by hanging. Sometimes the Governor-General reduces the sentence to 100 lashes, the ball and chain and detention for those on probation with an indefinite term. If a murderer is sentenced to death, the sentence is very seldom commuted. 'I hang murderers!' the Governor-General told me.

On the day before an execution, during the evening and throughout the entire night, the prisoner is prepared for his last journey by a priest. The preparation consists of confession and conversation. One priest told me:

At the beginning of my priestly career, when I was only twenty-five, I was ordered to prepare two convicts for death at the Voyevodsk prison. They were to be hanged for murdering a settler for 1 rouble 40 kopecks. I went into their cell. The task was a new one for me, and I was frightened. I asked the sentry not to close the door and to stand outside. They said, 'Don't be afraid, little father. We won't kill you. Sit down.'

I asked where I should sit and they pointed to the plank bed. I sat down on a water barrel and then gaining courage, I sat on the plank bed between the two criminals. I asked

what *guberniya* they came from and other questions, and
then I began to prepare them for death. While they were
confessing I looked up and saw the men carrying the beams
and all the other necessities for the gallows. They were
passing just below the window.

'What is that?' the prisoners asked.

'They're probably building something for the warden,' I
said.

'No, little father, they're going to hang us. What do you
say, little father, do you think we could have some vodka?'

'I don't know,' I said. 'I'll go and ask.'

I went to Colonel L. and told him the prisoners wanted
a drink. The colonel gave me a bottle, and so that no one
should know about it, he ordered the turnkey to remove the
sentry. I obtained a whisky glass from a guard and returned
to the cell. I poured out a glass of vodka.

'No, little father,' they said. 'You drink first, or we won't
have any.'

I had to drink a jigger, but there was no snack to go
with it.

'Well,' they said, 'the vodka brightened our thoughts.'

After this I continued their preparation. I spoke with
them an hour, and then another. Suddenly there was the
command: 'Bring them out!'

After they were hanged, I was afraid to enter a dark room
for a long time.

The fear of death and the conditions under which
executions are carried out have an oppressive effect on
those sentenced to death. On Sakhalin there has not
been a single case where the condemned man went to

his death courageously. When the convict Chernosheya, the murderer of the shopkeeper Nikitin, was being taken from Alexandrovsk to Dué before execution, he suffered bladder spasms. He would suffer a spasm and have to stop. One of the accessories in the crime, Kinzhalov, went mad. Before the execution they were clothed in a shroud and the death sentence was read out. One of the condemned men fainted when the death sentence was being read out. After Pazhukin, the youngest murderer, had been dressed in his shroud and the death sentence was read out, it was announced that his sentence had been commuted. How much this man lived through during that brief space of time! The long conversation with the priests at night, the ceremony of confession, the half-jigger of vodka at dawn, the words 'Bring them out,' and then the shroud, and then listening to the death sentence, and all this followed by the joy of commutation. Immediately after his friends were executed, he received 100 lashes, and after the fifth stroke he fell in a dead faint, and then he was chained to an iron ball.

Eleven men were sentenced to death for the murder of some Ainus in the Korsakov district. None of the officers and officials slept on the night before the execution; they visited each other and drank tea. There was a general feeling of exhaustion; nobody found a comfortable place to rest in. Two of the condemned men poisoned themselves with wolfsbane – a tremendous embarrassment to the military officials responsible for the execution of the sentences. The district commander heard a tumult during the night and was then informed that the two prisoners had poisoned themselves. When everyone had gathered around the scaffold just before the execution, the district commander found himself saying to the officer in charge:

'Eleven were sentenced to death, but I see only nine here. Where are the other two?'

Instead of replying in the same official manner, the officer in charge said in a low, nervous voice:

'Why don't you hang me! Hang me . . .'

It was an early October morning, grey, cold and dark. The faces of the prisoners were yellow with fear and their hair was waving lightly. An official read out the death sentence, trembling with nervousness and stuttering because he could not see well. The priest, dressed in black vestments, presented the Cross for all nine to kiss, and then turned to the district commander, whispering:

'For God's sake, let me go, I can't . . .'

The procedure is a long one. Each man must be dressed in a shroud and led to the scaffold. When they finally hanged the nine men, there was 'an entire bouquet' hanging in the air – these were the words of the district commander as he described the execution to me. When the bodies were lifted down the doctors found that one was still alive.

This incident had a peculiar significance. Everyone in the prison, all those who knew the innermost secrets of the crimes committed by the inmates, the hangman and his assistants – all of them knew he was alive because he was innocent of the crime for which he was being hanged.

'They hanged him a second time,' the district commander concluded his story. 'Later I could not sleep for a whole month.'

THE STORY OF PENGUIN CLASSICS

Before 1946 ...'Classics' are mainly the domain of academics and students, without readable editions for everyone else. This all changes when a little-known classicist, E. V. Rieu, presents Penguin founder Allen Lane with the translation of Homer's *Odyssey* that he has been working on and reading to his wife Nelly in his spare time.

1946 *The Odyssey* becomes the first Penguin Classic published, and promptly sells three million copies. Suddenly, classic books are no longer for the privileged few.

1950s Rieu, now series editor, turns to professional writers for the best modern, readable translations, including Dorothy L. Sayers's *Inferno* and Robert Graves's *The Twelve Caesars*, which revives the salacious original.

1960s The Classics are given the distinctive black jackets that have remained a constant throughout the series's various looks. Rieu retires in 1964, hailing the Penguin Classics list as 'the greatest educative force of the 20th century'.

1970s A new generation of translators arrives to swell the Penguin Classics ranks, and the list grows to encompass more philosophy, religion, science, history and politics.

1980s The Penguin American Library joins the Classics stable, with titles such as *The Last of the Mohicans* safeguarded. Penguin Classics now offers the most comprehensive library of world literature available.

1990s The launch of Penguin Audiobooks brings the classics to a listening audience for the first time, and in 1999 the launch of the Penguin Classics website takes them online to a larger global readership than ever before.

The 21st Century Penguin Classics are rejacketed for the first time in nearly twenty years. This world famous series now consists of more than 1300 titles, making the widest range of the best books ever written available to millions – and constantly redefining the meaning of what makes a 'classic'.

The Odyssey continues ...

The best books ever written

PENGUIN 🐧 CLASSICS

SINCE 1946